T0202406

Artificial Era

Artificial Era

Predictions, Problems, and Diversity in AI

GISSEL VELARDE

OXFORD
UNIVERSITY PRESS

Great Clarendon Street, Oxford, OX2 6DP,
United Kingdom

Oxford University Press is a department of the University of Oxford.
It furthers the University's objective of excellence in research, scholarship,
and education by publishing worldwide. Oxford is a registered trade mark of
Oxford University Press in the UK and in certain other countries

Published in the United States of America by Oxford University Press
198 Madison Avenue, New York, NY 10016, United States of America

British Library Cataloguing in Publication Data
Data available

Library of Congress Control Number: 2022950960

ISBN 978-0-19-286977-7

DOI: 10.1093/oso/9780192869777.001.0001

Printed and bound by
CPI Group (UK) Ltd, Croydon, CR0 4YY

Praise for Artificial Era

"Original, informative and easy to read. A good guide for understanding the future and facing it".

•

"This book gives a general introduction to the topic of Artificial Intelligence, making a deep analysis with a very solid scientific basis, without being too technical. It shows the current situation and the possible evolutions of AI, giving an objective and grounded position, highly welcome in these times of scientific misinformation. Highly recommended for the novice and the expert willing to broaden their perspective".

•

"A very natural and human vision of this new Artificial Era".

•

"There are already numerous books on artificial intelligence and its social impact, but Gissel Velarde's book has two characteristics that make it valuable and different. The first is that it is based on some 300 references in its bibliography, which gives it a very remarkable scientific character. The second is that AI is presented from the point of view of a Bolivian woman, who has lived in several European countries, and with a multidisciplinary professional profile. Thus the book, originally written in Spanish, conveys a diverse view of AI that addresses topics such as the role of women, technology in remote places, and references Potosi, India, the arts, drawing on studies of diverse cultures and geographies".

Emilia Gómez
AI and Music researcher
emiliagomez.com

•

"Based on her years of experience both in academia and in industry, as well as her interest in the world of business, Dr. Velarde presents in this book a realistic perspective of the role technology, in particular artificial intelligence (AI), is playing and will play in our lives both at a personal level and at the society level. Instead of focusing only on the potential dilemmas of general artificial intelligence, she discusses important topics including the need for national and international strategies for AI development, as well as the consequences of developing biased AI models in a world with large inequalities (gender, racial, class, etc.)".

Carlos Cancino-Chacón
Assistant Professor
Institute of Computational Perception, Johannes Kepler University
Linz

•

"From the very beginning, the book invites us to think, to reflect, to question; and it does it from the freedom that we have to position ourselves in some place of the world of knowledge and reasoning. Is it fiction? Is it reality? For those of us who have experienced the Era of the rotary phone, the first computer, the first cell phone, the first tablet; the book helps us to be aware of the dimensions of the great technological and social transformations that the former are capable of generating, but in addition to visualizing even greater ones. But at the same time, it alerts us to prepare our actions, oriented to build a social and human transformation towards a welfare state for the human race. As a humanist, lover of the arts, Gissel does not abandon her essence of putting the human being at the center of everything. Is it possible to humanize an Era marked by the artificial? What is the human origin and purpose of artificial ones? How does the artificial allow us to close, rather than widen, the infamous human gaps? I invite you to read this text without fear or prejudice, enjoy it from beginning to end not only to include it in the reading list of the year, but to reflect, decide and act".

Willy Castro Guzmán
University Professor and Researcher
Universidad Nacional, Costa Rica

•

"With a simple and precise language, the author talks about Artificial Intelligence (IA) and its subareas—machine learning and deep learning—through practical examples, graphs and the collection of scientific studies on the subject along the years. Velarde goes through topics ranging from philosophy, creativity, consciousness, to of course, intelligence and education. All this, taking as a central axis AI and its impact worldwide and in Latin America".

Dafne García
Journalist
LadoBe

•

Contents

•

•

Contents

Read critically and at your own risk.

Contents

List of Figures

List of Tables

Prelude

In an Artificial Cataclysm, which species would survive? And of these species, which would dominate the Earth? Would this be a more collaborative or a more tyrant one? From the cataclysm that occurred some 65 million years ago, it is known that the great dinosaurs disappeared, but organisms able to adapt to the extreme conditions of the planet survived, such as insects, birds, and reptiles. We arose from the lucky species. Favorable conditions, luck, and the development of our cognitive abilities made us the sovereign beings of this Era. We are more innovative, complex, social, and fearsome than the most imposing and ruthless dinosaurs. Also, we consider ourselves the most intelligent species on Earth, although perhaps our greatest merit is to transmit our culture efficiently over time.

We have gone from trees to skyscrapers, and from walking to being driven by autonomous vehicles; from spear hunting to developing agriculture and industry. Language has allowed us to accumulate collective knowledge, and we have evolved to enjoy music. We are increasingly extending our life expectancy. The barriers of time and space to work, learn, and socialize in real-time do not exist for telecommunications and for those who can enjoy this fantasy.

There is no shortage of stories of those who with a brilliant idea and computational skills, have gone from a garage to an empire in the Artificial Era in just a few years. Data is the new gold that users give to data miners. Information reaches more corners of the planet and we will enjoy more innovations than several past generations combined. The technological surrealism driven by intelligent automation seems to lead to fierce competition with even more comfortable lifestyles for those who can afford them. But if life would come down to a Go board, we could all celebrate retirement, or return to slavery, and even disappear from the face of the Earth. Artificial intelligence deserves special attention because of its power and versatility.

Is it possible to be prepared for the unexpected? The giants will wonder how to avoid being overshadowed by the little ones, perhaps developing collaborative strategies. And the little ones will have to be careful not to die trying. In the end, if the game would abruptly change the luck of the players, new rules could make it more interesting for everyone. To avoid disastrous conflicts or face existential dangers, global collaboration seems to be the way to go.

For about twenty years, I have dedicated myself to the serious study of the computational world. Over time, I discovered that it was more difficult for me to understand people than computers, but understanding both and their interaction is essential, especially now. In this book, I try to understand and tell a story about the Artificial Era from the lens of a Latina woman, who has seen two worlds. Although perhaps she may be bold or naive and still has much to learn, she was guided along the roads of excellence and has been able to teach and put into practice what she has learned, with successes and failures. I believe that my perspective can be enriching not only for Artificial Intelligence communities, but for all who are affected by intelligent automation.

Writing this book has been like an adventure inside a maze of corridors with doors that, when opened, lead to more mazes, corridors, and doors. I have been able to deepen my knowledge in the fields I master, but also know that it was risky to enter lands far from my specialty. Besides, my story could sound biased; it is difficult to escape from one's bubble, and sometimes it is necessary to push in a certain direction to stabilize the balance. I hope this book will be useful to the individuals who can make a difference in these intense times.

Chapter 1

The Artificial Cataclysm

The news of human defeat by the machine fell like the impact of a bolide meteor at the center of human ego. The intelligent machine would have felt neither pleasure nor compassion at seeing the former world Go champion bow to its strategy. At least, not yet. The news spread in seconds like a tsunami with gigantic waves lasting for months at nonsubsiding pulses. Although it was not the first time that artificial intelligence subdued human experts in their respective fields, this time was different. Political leaders opened their eyes like full moons and rubbed their hands. The race became serious. Institutional plans and budgets to master the technology ignited like erupting volcanoes. Researchers knew that the technology had matured as never before and had known it for a few years, but most of them did not come down from their scientific cloud. Hurricanes of interesting, disturbing, and curious news began to spread at breakneck speed, which kept people attached to their smartphones. Never before had the news been so surreal. Investors began to calculate and recalculate their financial projections. Their eyes glowed monetary symbols, though some were unaware of the past winters. Nonetheless, during the news storm, the breezes brought to our ears the song of a choir of sirens from the high seas announcing the arrival of a promised technological spring.

The fantastic predictions did not come without warnings. Some had horrible nightmares with millions of people in despair due to the loss of their jobs. The exodus of these unemployed was then efficiently executed by autonomous vehicles, robotaxis. Families in procession entered old houses as new homes, in possession of a few suitcases. The thriving companies no longer knew what to do with all the income that robotic secretaries, nurses, hoteliers, or bricklayers provided and all without rest or unionization. Many millionaires decided to burn their fortunes in capricious projects classified under the Omega label. During troubled dreams, smartphones became an extension of people's bodies. People disconsolately wondered why they were more obedient, more creative, smarter, and less complicated than they were. Many threw themselves off hanging bridges because their lives ceased to have meaning. Drenched in a cold sweat, the nightmare dreamers awoke.

Others, however, gathered to celebrate the arrival of a new chapter. They celebrated the new capabilities of robots, as these could finally perform tasks previously trusted to careful workers' sensory perception and motor skills. Robots possessed skills to look, assemble parts, select defective products from good ones, pack them in boxes, and ship them through a logistics chain to which human intervention would only cause delays. Other individuals enthusiastically told happy stories about their investments in technology companies that were growing in record time. Outside robotic factories, rivers of artisans, merchants, and the multicraft were downloading and seamlessly installing the intuitive applications of Internet platforms to work and finally enter a system that allowed them to easily pay their taxes. The conversations acknowledged that the new force was not one of creative destruction but rather an organizing force.

Those who celebrated also spoke of new centers of innovation, where appropriate policies allowed the emergence of dynamic and diverse cultures of entrepreneurs, scientists, investors, and all kinds of professionals and workers, creating innovations that shaped millions of people's lives. With a palette of inventions, people became more and more productive. Some of those innovations became million-dollar businesses, while others entered the whirlwind of creation and destruction. Many saw products and services come and go from all over the world, even from the most remote places. This international competition was threatening to some because of the speed at which things were happening. The formerly vulnerable managed to catch up with technology. They were all testing their ability to resist and adapt to the new conditions. Artificial intelligence was still in fashion and progressively entered every industry and sector, merging with other incredible technologies. There was talk of new gold in circulation. There was also talk of a generation of ultrahumans. Some diseases were relegated to history books. Nevertheless, end of the world predictions continued playing like the ostinato violins of an action movie. And, to this theme, a suspended chord that announced the end of the human species in the distance would be added. Meanwhile, the news of a new world pact was heard, and the oracles wept with emotion. They also cried because they knew the transformation could be painful, with several possible outcomes.

If we really are in the middle of a technological revolution similar to an Artificial Cataclysm, who will benefit the most and who will be the unlucky ones? Why have several countries invested in an artificial intelligence strategy? Which strategy should people and organizations follow to harness the power of the smart wave? Is a wonderful world waiting for us, or a disastrous one? Have algorithms become smarter than humans? Who are the ultrahumans? Will robots ever become conscious, and if so, will they help us or tear us apart? Let's take it one step at a time. First, we will place ourselves in time to know where we are.

The Surreal Technological Progress

Although we are living in tough times, a terrific and unrepeatable opportunity is coming. Experts are expecting an event of gigantic magnitude and profound repercussions, artificially caused by the actions of the current civilization. It is not the first time that humanity has gone through an Industrial Revolution or a Technological Revolution—both terms will be used interchangeably—but this revolution is loaded with various technologies with far-reaching implications. It could unleash scenarios so bizarre that even the most successful futurists would not imagine. It is difficult or rather impossible to imagine what one does not know or extrapolate one's experience into an artificial future.

What is known about previous technological revolutions is that they presented patterns. The next technological revolution begins shortly before the stage of maturity and end of the previous one; like a wave that has a beginning and a stage of growth until it breaks with full force. Anyone who has ever surfed or seen others surf knows that depending on the location, surfers could lose the wave by not taking advantage of its strength, or they could ride it if well prepared, but also could fall and even die under the immensity of the sea.

We are at the precise moment when a surfer should catch the wave to take advantage of its strength. Where does this idea come from? It is what scholars like Carlota Perez, Klaus Schwab, or Spyros Makridakis expect because there is sufficient historical evidence from previous industrial revolutions. Technological waves resemble the waves of the sea, which steer us towards growing Technological Progress. If we do not catch the approaching wave, we will see it pass by in front of our eyes, and there will not be another similar opportunity for many years. It could even be the last one we see. These waves last for about half a century, and there is always the remote possibility of an absolute disaster.

Researchers agree that the greatest beneficiaries of Technological Progress are generally those who acquire key technologies at the beginning of the industrial revolution in which they are found. While Klaus Schwab and Spyros Makridakis consider us in a fourth technological wave, Carlota Perez states that we are already in the middle of the fifth. In any case, they all agree that we are at a critical point of the most spectacular industrial, technological revolution ever seen.

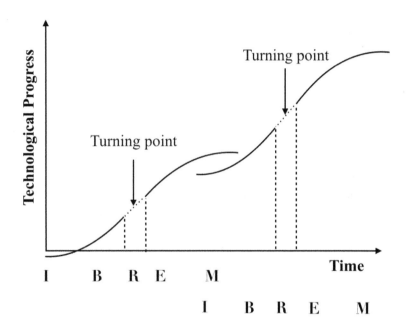

Figure 1.1: Technological Progress over time; two IBREM-type patterns are observed with stages of Irruption, prosperity Bubble, Recession, golden Epoch, and Maturity. The following pattern begins before the end of the previous pattern. Based on Perez (2003, 2016, 2019).

Carlota Perez studied patterns of the technological revolutions and observed that after a period of technological irruption, a prosperity bubble occurs, followed by a period characterized by hard spans of recession and panic that can last for several years, during which there is a turning point before an epoch of great prosperity, and finally, a period of maturity, as shown in Figure 1.1.

From the beginning of the eighteenth century until now, in a period of a little more than three centuries, Perez found that four and a half technological waves occurred under the pattern of Irruption, prosperity Bubble, Recession, golden Epoch, and Maturity (IBREM) as seen in Figure 1.2. According to Perez, we are currently in a period of recession, which has had periods of recovery and relapse and could continue for several years, with even unpleasant events occurring if good global policies are not applied, and a proper global lifestyle is not adopted.

The effects of technological revolutions on the world population have been significant. The world's population has remained in the millions for several millennia. According to Worldometers, there were approximately 610 million inhabitants on Earth in 1700 and 1 000 million by 1800. In 1930, the world population doubled and only 30 years later, it tripled such that in 1974, the highest annual population growth rate was reached. Since then, the growth rate has been decreasing. In 2020, the planet had 7 800 million people, and Worldometers estimates that the world population will reach 9 735 million inhabitants by 2050.

The First Industrial Revolution marked a milestone when the steam engine and other machinery were introduced to perform an endless number of activities without relying on human strength. In addition, the first vaccine was developed. The Second Technological Revolution brought significant advances with electricity, automobiles, railroads, radio, and penicillin, which are all still vigorously used to this day. The Third Technological Revolution brought continuous production lines and worldwide communication through telegraphy. The Digital Revolution also brought several inventions such as general-purpose computers, the Internet, and smartphones at affordable prices with which it is possible to perform an end-

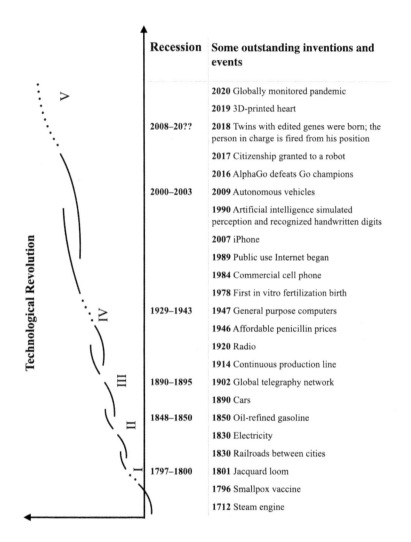

	Recession	Some outstanding inventions and events
V		
		2020 Globally monitored pandemic
		2019 3D-printed heart
	2008–20??	**2018** Twins with edited genes were born; the person in charge is fired from his position
		2017 Citizenship granted to a robot
		2016 AlphaGo defeats Go champions
	2000–2003	**2009** Autonomous vehicles
		1990 Artificial intelligence simulated perception and recognized handwritten digits
		2007 iPhone
		1989 Public use Internet began
		1984 Commercial cell phone
		1978 First in vitro fertilization birth
IV	**1929–1943**	**1947** General purpose computers
		1946 Affordable penicillin prices
		1920 Radio
		1914 Continuous production line
III	**1890–1895**	**1902** Global telegraphy network
		1890 Cars
II	**1848–1850**	**1850** Oil-refined gasoline
		1830 Electricity
		1830 Railroads between cities
I	**1797–1800**	**1801** Jacquard loom
		1796 Smallpox vaccine
		1712 Steam engine

Figure 1.2: Technological Revolutions, some remarkable inventions and events. Based on various sources.

less number of activities: communication, selling, buying, working, studying, or playing.

There are several who consider Artificial Intelligence (AI) to be the omnipresent force that will drive this technological revolution. AI became a favorite daughter of investors. The consulting firm PricewaterhouseCoopers estimated that by 2030 AI will contribute some USD 15.7 trillion to the global economy. Then Accenture consultants, Purdy and Daugherty, estimated that by 2035, AI would double the world's economic growth rates; first, by making workers 40 percent more productive; second, by becoming a virtual workforce; and third, by empowering endless innovations creating new revenue streams.

Andrew Ng compares AI to electricity as he thinks AI will be widely adopted and will transform all major industries. Spyros Makridakis calls this revolution "the Artificial Intelligence Revolution" and predicts that by 2037 we will be living in a climactic period. Additionally, Klaus Schwab says in his 2016 book that there are three megatrends for this revolution, which are:

- first, physical innovations like robots of all sizes and for all kinds of applications (from nanorobots to robotic factories), autonomous vehicles, 3D printing, and new materials,

- second, digital innovations that will harness communications for greater connectivity and will provide endless applications and platforms, and

- third, biological innovations, which promise better health.

Although when talking about surrealism, we could think of the works by Salvador Dalí, which capture the impossible in the possible, and mix reality with fantasy; we can also think that life today, its extreme contrasts and the latest advances in science and technology, show us that everything is possible, inexplicable and incredible. In 2015, Samuel Sánchez said that soon nanorobots, ones much smaller than the diameter of a hair, are expected to travel intelligently through the bloodstream to efficiently combat diseases such as cancer. Besides, in the not too distant future, it will be possible to create tissues and organs by 3D printing them with the same cells of those who will wear these biotechnology masterpieces. In 2019, AFP Español presented a report showing the prototype of a 3D–printed heart developed by Nadav Noor, Tal Dvir, and their team at the laboratory of Tel Aviv University.

A human colony will travel six months to land on Mars, as planned by the SpaceX company, making Mars the next habitable planet closest to Earth in which gold, agate, and geodes could exist. Meanwhile, others could live in floating, modular marine colonies assembled like LEGO and created to fulfill the 17 Sustainable Development Goals of the United Nations, especially in addressing the threat of rising seas.

A mason robot can autonomously build a three-bedroom, two-bathroom house in less than three days. A nurse robot can entertain more than 20 elderly people with games and group exercises. There are other robots to provide customer services in hotels and restaurants. There are English teaching assistant robots, which are operating in 500 classrooms in Japanese schools.

Simultaneously to all this colossal fantasy of surrealistic progressive technology, there are other realities. In the USA, the same country where SpaceX is working to offer trips to Mars to a multi-planetary society, a 2020 article by Reuters Staff claimed that non-Hispanic White families are 10 times richer on average than Black families. Other statistics include that a Black male is much more likely to go to jail than a White male, or a Black woman is more likely to die from pregnancy-related causes than a White woman.

Germany was led for many years by "the most powerful woman in the world", a title awarded to Angela Merkel by Forbes magazine between 2005 and 2021. At the same time, her country presented one of the largest gender pay gaps in Europe. According to a 2019 European Commission report, the highest-paid positions generally within science, technology, engineering, and mathematics fields are more than 80 percent filled by men, then management positions by 90 percent, and with salaries 23 percent higher than those of the women who barely make it to the top of the ladder, like their ex-leader.

In 2010, the World Bank estimated that the regions of low-income countries in the Middle East, Africa, Eastern Europe, Asia, Latin America and the Caribbean, which contribute little to CO_2 emissions and where a large part of their population relies on agriculture, will have to deal with increases in droughts, floods, and forest fires. Yet, high-income countries whose industry, nonalternative energy, deforestation, and industrial agriculture account for the largest percentage of CO_2 emissions, will be the least affected by natural events caused by climate change. If the experts' predictions are correct, in a century, the sea level could rise by enough meters for several regions to be flooded and 40 percent of the world's population to be displaced.

One technological innovation that has caused an international stir is gene editing. Gene editing allows for changes to be made in the genetic code of living organisms so that their nature is modified. In 2020, researchers Jennifer Doudna and Emmanuelle Charpentier received the Nobel Prize in Chemistry for their contribution to this field of research. Jennifer Doudna explained in a 2019 presentation that gene editing can be used as a nonhereditary treatment to affect a single person or as an inherited treatment to affect the treated individual and their descendants.

Gene editing is viewed by some as "playing God". In 2018, in the middle of the experts' delicate debate over gene editing's far-reaching implications, two twins with edited genes were born. The genetic research committees were reportedly in shock upon hearing the news. Although the person responsible for the experiment was fired from their position, the experiment cannot be undone. Its secondary consequences will not be known for some time.

To all of this, we have to add the possibility of artificial superintelligence's arrival; since once again, artificial intelligence has shown

Figure 1.3: Human playing Go; sketch by the author.

itself superior to our brightest humans, this time in the strategy game of Go.

The Stopwatch has Started

Already by 1997, artificial intelligence proved itself by beating the world chess champion. In 2011, it defeated the best players in the general culture question game called Jeopardy. Then in 2016, the overwhelming victory of AlphaGo's algorithm over Lee Sedol, the South Korean until then world champion of the ancient strategy game of Go, unleashed existential questions about the limits of human intelligence against a superior artificial entity.

Go players claim to experience a contemplative state of mind when playing this game. They believe they explore the boundaries of human thought by developing game strategies between two opponents, usually on a 19 by 19 line board, filled with black and white stones to surround the most possible territory. For computers, it was simply an impossible challenge to solve by traditional search methods due to the number of possibilities to calculate.

AlphaGo would have felt no euphoria nor frustration at winning or losing in the 2016 matches against Lee Sedol. The film *AlphaGo* documents the preparation and historical matches in which Lee seemed to have played representing humanity against a crushing artificial intelligence created by a team of 20 researchers. So one could think that the matches were 1 against 21. More than one observer must have been moved as they watched Lee's ordeal playing out in front of the machine.

In the years following the 2016 matches, DeepMind's team continued developing new and more powerful versions of its artificial intelligence algorithms dedicated to Go. Reporter James Vincent announced that Lee Sedol would retire from the professional Go league in 2019, because even though he is the number one human in the world, for him, there is an invincible entity.

In 2017, Vladimir Putin said that whoever leads in artificial intelligence "will dominate the world". From that year until now, several countries have made an effort to develop, allocate significant budgets, and implement their artificial intelligence strategies. Canada was the first to make its strategy known, followed by Japan, Singapore, China, the United Arab Emirates, Finland, Taiwan, Denmark, France, the United Kingdom, Australia, USA, South Korea, Sweden, India, Mexico, among others. Various government initiatives are being executed, such as the creation of laboratories, research centers, master's degrees, or artificial intelligence expert groups, as we will see later on. International alliances have also been formed by the United Nations, the European Commission, the Nordic-Baltic Region, the leaders of the G7, the International Study Group on Artificial Intelligence and the agreement between the United Arab Emirates and India. The issues that arise for these alliances often focus on education, labor skills, social benefit, economic growth, competitiveness and cooperation, ethics and regulation, research, innovation, and digital infrastructure.

The interest in artificial intelligence is evident. Further on, we will examine the technology trends in theoretical and applied research to get an idea of its development worldwide. But, have algorithms really become more intelligent than humans? How can we know if an algorithm is intelligent? And what about jobs? Will robots steal all our jobs or will they make us more productive? First of all, it is worth knowing how this field of research developed.

The Journey into the Deep

Human beings have always sought out intelligent and autonomous entities to employ and free themselves from some tasks, so that they can concentrate on what they want the most. Aristotle, 384–322 Before the Common Era, thought that servants and slaves would cease to exist, if every possession we had could operate autonomously and intelligently to our desire or sense of need. For example, Aristotle imagined statues moving by themselves and lyres playing their own music.

In the mid-1850s, Ada Lovelace, recognized as the first programmer in history, thought that analytical machines, a mechanical prototype of digital computers, could be used to solve all kinds of complex problems, such as automatically composing elaborate music. But it took almost a century for artificial intelligence to begin to take shape. In 1950, Alan Turing rescued Lovelace's work, proposed a "logical computing machine (LCM)", today called a "Turing machine", considering that it was possible to build general-purpose computers. He eventually posed the famous question of whether or not machines can think, a question that keeps many awake at night and beyond. He also proposed how to evaluate artificial intelligence through a test that bears his name.

Nils Nilsson claims that artificial intelligence's field of research was consolidated with three scientific events held in 1955, 1956 and 1958: *Session on Learning Machines, Summer Research Project on Artificial Intelligence, Mechanization of Thought Processes*, respectively. Soon, intelligent algorithms dazzled the world with their ability to solve problems intellectually difficult for humans, such as advanced geometry exercises and mathematical theorems. Nonetheless, they failed miserably at tasks that babies do naturally, such as recognizing faces or performing motor tasks. This paradox was identified by Hans Moravec around 1988, and haunted artificial intelligence like a ghost for several years.

To mimic the brain's processing in perceptual tasks, such as face recognition, it was necessary to study its functioning from various research fields. Thanks to Santiago Ramon y Cajal and Camillo Golgi, who received a Nobel Prize in 1906, we know that neurons and their connections are responsible for brain function. In 1962, David Hubel and Torsten Wiesel discovered groups of neurons activate from specific stimuli and hierarchically pass information from one layer to the other in a process of abstraction. Convolution, a mathematical tool described in its general form in 1754 by Sylvestre François Lacroix, was used in different research centers in 1980 by Stjepan Marčelja and John Daugman, demonstrating resemblance to neural activation by signal filtering, thus having implications on pattern recognition.

With the advent of general-purpose programmable computers in 1947, it became possible to program, test, and develop new algorithms. Frank Rosenblatt's Perceptron, initiated the concept of an Artificial Neural Network in 1957. Although Artificial Neural Networks were seen as promising models for imitating brain function, they stayed confined to the laboratory for many years as they relied on hardware developments needed to perform their heavy computations. Meanwhile, other algorithms, such as Support Vector Machines combined with various types of data representations, would become a popular choice.

Researchers were determined to find robust representations that would allow them to mimic perceptual tasks. Finally in 1990, Yann LeCun and his colleagues developed a layered, hierarchical Convolutional Neural Network, which demonstrated being able to classify

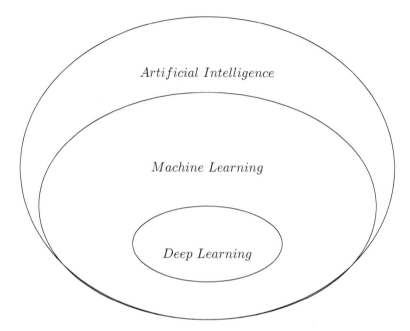

Artificial Intelligence

Machine Learning

Deep Learning

Figure 1.4: Artificial intelligence and its most representative subareas, Machine Learning and Deep Learning. Based on I. Goodfellow et al. (2016) and World Intellectual Property Organization (2019).

accurately handwritten digits from letter zip codes; a task previously possible only by human visual inspection. Years later, these hierarchical processing models, capable of simulating some brain processes, were given the name *deep learning*. Global attention would not come until 2012, when Alex Krizhevsky, Ilya Sutskever, and Geoffrey Hinton implemented a deep convolutional model trained by a large volume of digital images using powerful computers called Graphic Processing Units (GPUs). In the years that followed, the artificial intelligence community was fascinated with deep learning. The straw that broke the camel's back hit in 2016, when David Silver's DeepMind team developed the AlphaGo algorithm that became the undisputed Go champion and challenged humanity in its strategic and creative capacity. This last event deeply infiltrated the news. Artificial intelligence has been portrayed in the media as the promise and downfall of humanity.

Artificial intelligence's path has not been a triumphant one the whole time. Rather, it is better illustrated as a roller coaster ride with acceleration, great climbs and noisy descents called the Artificial Intelligence "Winters". Currently, research in this field is experiencing a splendid time. Artificial intelligence encompasses machine learning, and this in turn, encompasses deep learning as represented in Figure 1.4.

Empowerment from artificial intelligence in all technological inventions is growing rapidly. This can be seen by the number of AI-based patent applications. Eighty-nine percent of artificial intelligence patents use machine learning, and deep learning patent applications have grown on average by 175 percent annually between 2013 and 2016. We will go further into detail later.

Intelligence and Artificial Intelligence

Here a collection of definitions of intelligence and artificial intelligence. According to neurologist Richard Haier, intelligence has to do with the brain and in his 2016 book he defines:

> Intelligence is a *general* mental ability" [...] "Intelligence is what we call individual differences in learning, memory, and attention.

Another definition of intelligence, also mentioned by Hair as widely accepted in research, is by psychologist Linda Gottfredson from 1997:

> Intelligence is a very general mental capability that, among other things, involves the ability to reason, plan, solve problems, think abstractly, comprehend complex ideas, learn quickly and learn from experience. It is

not merely book learning, a narrow academic skill, or test-taking smarts. Rather, it reflects a broader and deeper capability of comprehending our surroundings— "catching on", "making sense" of things, or "figuring out" what to do.

In 2010, Nils Nilsson, a veteran of computer science, declared:

> For me, artificial intelligence is that activity devoted to making machines intelligent, and intelligence is that quality that enables an entity to function appropriately and with foresight in its environment.

Patrick Winston, another veteran of computer science, stated in 1992:

> Artificial intelligence is ... The study of the computations that make it possible to perceive, reason, and act.

Machine learning algorithms learn from data and through experience without the need for detailed instructions. This was demonstrated by Arthur Samuel in 1959, with his development of an algorithm that learned to play checkers.

After the artificial intelligence winters, the label *artificial intelligence* or *AI* ceased to be used for a while. The label *data mining*, which emphasizes the extraction of knowledge from data, was quite popular in the business world and gained prominence in the early 2000s. The term *data miner* was adopted to refer to the person in charge of data mining. Today they can also be called *data scientists*. The label *machine learning* describes primarily algorithms that learn from data and would become relevant after 2010. But people dedicated to machine learning are not called *machine learners*; they are generally called *machine learning scientists* or *engineers*. Another term that took prominence after 2012 was *deep learning*, and as seen in Figure 1.5, the *AI* and *artificial intelligence* labels increased in popularity again.

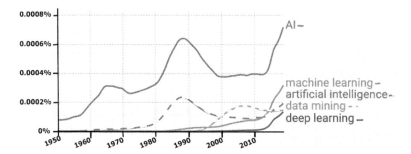

Figure 1.5: Frequency of terms: AI, machine learning, artificial intelligence, data mining, and deep learning between 1950 and 2019; Based on data from Google Books Ngram Viewer (n.d.). It is possible that the abbreviation AI is related not only to artificial intelligence, but also to other abbreviations not related to artificial intelligence.

Sometimes, I have the impression that people think that artificial intelligence has "magical" powers. There is no doubt the scientific community is proud and enthusiastic about the achievements in the field, but logically, not everyone understands in depth what AI is all about. Intuitively, I think that the label *artificial intelligence* is better marketed than, for example, the label *machine learning*. Besides, it promotes discussion about what is understood by intelligence, making us review the concepts of intelligence. The label *AI* also incites us to fantasize about the possibility that algorithms really could simulate brain processes, but many times, intelligent algorithms simply solve a problem even as a black box to which we do not understand very well.

Today, intelligent machines possess some human capabilities, and additional ones that can be exploited to enhance human capabilities of perception, creativity, memory, pattern finding, precision, and thus increase labor productivity, creativity, improve decision-making, or simply automate certain tasks, all to help us obtain an augmented capacity, as shown schematically in Figure 1.6.

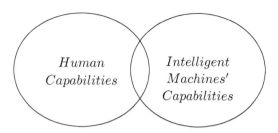

Figure 1.6: Augmented capacity and relationship between human capabilities and the capabilities of intelligent machines.

Machine Learning

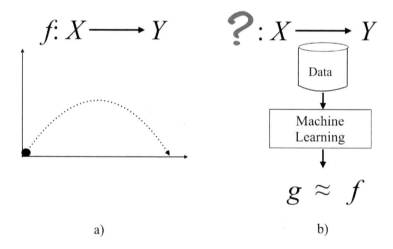

a)   b)

Figure 1.7: a) Function f is known and machine learning would not be needed to predict where a projectile will fall, if we know the direction and the speed with which it is shot. b) Function f is unknown, but it can be approximated by supervised machine learning if we have data (examples X and its labels Y), such that $g : X \to Y$ approximates f.

Algorithms are sequences of steps that solve a problem. Examples of algorithms are recipes, the sequence of actions that one takes to go to the movies, or the sequence of steps to find a pattern. Machine learning is about finding solutions based on the available data through a process of generalization. If we cannot describe a phenomenon with the help of mathematics, but there are patterns in the data we have, machine learning can help us find approximate solutions automatically.

In a game with slings and stones, you can predict where the fired stones will fall by simply using the equations of Newton's laws; that is, if you know the direction and speed with which each stone is fired. If we have the mathematical model of a physical phenomenon, we would not need to use machine learning. However, there are many applications where we have no idea what the mathematical description is, but we have sufficient data recorded and machine learning can find a final hypothesis that approximates our target function, which is the riddle depicted in Figure 1.7.

Let's say we want to predict the next word written in an email, but we do not have a mathematical formula that tells us how to do that. Or, we want to automatically recognize gender out of thousands of face images, but we do not know how we do that. Or, we want to use a system to help us decide which client should be granted credit, but we do not have a credit decision formula. For all these examples, we can use machine learning to solve the problem with some certainty and error. Machine learning depends on the quality and quantity of data and the algorithms used to extract patterns from a dataset.

Quality and quantity of data are essential for intelligent algorithms. Without data, there is no machine learning. If you do not have a dataset, it is possible to create one, by means of an experiment. Yet, it is advisable to start working with available datasets. Many researchers discard projects if data is not available. It is also essential to make a good review of the existing literature on state-of-the-art algorithms. In recent years it has become very common for scientific papers to share their implementation of algorithms with references to the datasets used, allowing their research to be reproducible.

There are datasets of faces, urban scenes, music genres, text, reviews and ratings of products in e-commerce stores, credits, insurance, temperature measurements, X-ray images, patients and their lab tests plus their medication, spam emails, economic indicators, stock prices, pedestrians and traffic signs, and much more. Each dataset contains examples, also known as objects, instances or points, with characteristics (*features*), which describe them, for example, their heights, widths, colors, content, etc.

If we had digitized thousands of popular songs, we could apply machine learning to classify or catalog them automatically, so that when the recording of an unknown band appeared, an intelligent machine could catalog the song according to how it sounds, without the need for a music expert to assign it to a music genre. A music store could receive thousands unpublished recordings daily, and could organize their catalog simply by using intelligent machines to assign the recordings to their genre. If algorithms can find relevant patterns to distinguish between music genres, a logical step is to use algorithms to create music with what they have learned.

The *ground truth* of a dataset may reflect factual or subjective information. For example, the *genre* label of a music recording is assigned subjectively based on perceptual and cultural cues, while the *band* or *performer* label is factual information, unless it was a recording lacking authorship information, and our task was to find out who made the recording.

Some books that I find essential and recommend for anyone who wants to apply machine learning and deep learning are:

- *Introduction to Machine Learning* by Ethem Alpaydin,

- *Deep Learning* by I. J. Goodfellow, Bengio, and Courville,

- *Reinforcement Learning: An Introduction* by Suton and Barto,

- *Data Mining* by Witten, Frank, and Hall,

- *Learning from Data: A Short Course* by Abu-Mostafa, Magdon-Ismail, and Lin,

- *Pattern Recognition and Machine Learning* by Bishop, and

- *Combining Pattern Classifiers* by Kuncheva.

Trends in Intelligent Technologies

My literature review article published in 2019 indicated that artificial intelligence was becoming a ubiquitous technology. I decided to survey people engaged in artificial intelligence communities to find out their opinion regarding trends in the field. Additionally, I found a publication from the patent office, presented next.

Patents are instruments for the commercial exploitation of technology. The World Intellectual Property Organization published a report called: *Technology Trends 2019: Artificial Intelligence*, in which 27 renowned professionals contribute, representing academia, industry, trade unions and in various areas such as artificial intelligence, business, entrepreneurship, economics, philosophy, politics, bioinformatics, and futurism. The report's figures are clear: in recent years, artificial intelligence is used to enhance most technologies, and many of these patents can be applied in various industries. One-third of patents use machine learning, and its use has been growing since 2013 at an average rate of 28 percent per year. The remaining technologies that do not use artificial intelligence grew by an average of 10 percent from 2013 to 2016.

The WIPO report revealed that since the mid-1950s, more than a third of a million AI-based patents were filed, and more than 1.6 million scientific articles were published in this field. WIPO report shows that the trend in recent years is changing from theoretical research toward commercial exploitation of artificial intelligence inventions, given by the ratio between scientific publications and inventions from 8:1 in 2010 to 3:1 in 2016. To understand the devel-

Techniques used	Patent filings in 2016	AAGR %
Deep learning	2 399	175
Multitask learning	n.i.	49
Neural networks	6 506	46

Table 1.1: Artificial intelligence techniques used with the highest growth between 2013 and 2016, number of patent filings in 2016 and average annual growth rate of patent filings between 2013 and 2016 (AAGR %), n.i. means: no information. Based on data from World Intellectual Property Organization (2019).

opment of artificial intelligence in the world of patents, WIPO has considered three dimensions:

- techniques used,

- functional applications, and

- application fields.

These are shown in Tables 1.1, 1.2, and 1.3, respectively, which only include the fastest-growing categories between 2013 and 2016, and not all WIPO categories.

WIPO indicates that 89 percent of patents related to artificial intelligence are in the area of machine learning. Deep learning experienced the greatest average annual growth, at a spectacular 175 percent. Computer vision was the most popular functional application, while Transportation had the highest number of patent filings within application fields. See more details in Appendix A.1.

In addition to compiling information from WIPO, I conducted an online survey called *AI Trends* to know the opinion of artificial intelligence researchers. Most of the survey participants were invited in 2020 via the following communication channels: *Machine Learning News*, *Women in Machine Learning*, *ISMIR Community Announcements* and *Women in Music Information Retrieval*, and few researchers received a direct survey invitation. Of the 51 anonymous and voluntary responses received, I analyzed below 21 responses from

Functional applications	Subcategory	Filings in 2016	AAGR %
Computer vision		21 011	24
	Biometrics	>6 000	31
	Character recognition	>4 000	n.i.
	Scene understanding	<4 000	28
	Image and video segmentation	>2 000	n.i.
Natural Language Processing (NLP)		>5 000	n.i.
	Information extraction	>2 000	n.i.
	Semantics	>1 000	33
	Machine translation	<1 000	n.i.
	Sentiment analysis	~50	28
Speech processing		>5 000	n.i.
	Speech recognition	>3 000	12
	Speaker recognition	<2 000	12
	Speech-to-speech	n.i.	15
Robotics		>2 000	55
Control methods		n.i.	55

Table 1.2: Artificial intelligence functional applications with the highest growth between 2013 and 2016, subcategories, number of patent filings in 2016 and average annual growth rate of patent filings between 2013 and 2016 (AAGR %), n.i. means: no information. Based on data from World Intellectual Property Organization (2019).

Application fields	Subcategory	Filings in 2016	AAGR %
Transportation		8 764	33
	Autonomous vehicles	5 569	42
	Aerospace/avionics	1 813	67
Telecommunications		6 684	23
	Radio and television broadcasting	n.i.	17
	Computer networks/Internet	n.i.	17
Security		>4 100	n.i.
Life and medical sciences		4 112	12
	Medical informatics	n.i.	18
	Public health	n.i.	17
Personal devices, computing and HCI		3 977	11
	Affective computing	n.i.	37
Business		>2 000	n.i.
	E-commerce	n.i.	n.i.
	Enterprise computing	n.i.	n.i.
Industry and manufacturing		>2 000	n.i.
Document management and publishing		<2 000	n.i.
Networks		n.i.	n.i.
	Smart cities	n.i.	47
Agriculture		n.i.	32
Computing in government		n.i.	30
Banking and finance		n.i.	28

Table 1.3: Artificial intelligence application fields with the highest growth between 2013 and 2016, Subcategories, number of patent filings in 2016 and average annual growth rate of patent filings between 2013 and 2016 (AAGR %). The table shows 12 of the 20 categories of WIPO. n.i. means: no information. Based on data from World Intellectual Property Organization (2019).

those who identified themselves as professors and postdocs. Two postdocs were in Africa, four professors and three postdocs in America, four professors in Asia, and three professors and five postdocs in Europe. Five participants identified themselves as female, fourteen as male, one as nonbinary (genderqueer), and one as genderless. Appendix A.2 includes the invitation and the complete survey.

Figure 1.8a shows the word cloud generated based on the participants' answers to the question: "In your opinion, what are the Top 5 trends in AI research?" The words that stand out in the word cloud according to their font size are:

- *learning,*

- *AI, machine, deep learning, reinforcement,*

- *computer vision, NLP (natural language processing), ML (machine learning), GANs (generative adversarial networks),*

- *application, robotics, driving, systems, nonsense,*

- *ethics, fairness, problems, recognition, autonomous, health, planning, simple, speech, natural, language, processing, analysis, medical, recommender, reality.*

Instead, to the question: "What should be the Top 5 priorities in AI research?" the words that stand out the most, according to their size, are those that can be seen in the Figure 1.8b:

- *AI, learning,*

- *ethics, data,*

- *human, machine, deep, world,*

- *understanding, computer, vision, medical, processing, energy, healthcare, applications, system,*

- *big, change, climate, policy, automatic, intelligent, companies, analysis, transport, detection, robotics, interaction, understand, market, social.*

(a)

(b)

Figure 1.8: (a) Word cloud based on answers to the question: "In your opinion, what are the Top 5 trends in AI research?" (b) Word cloud based on answers to the question: "What should be the Top 5 priorities in AI research?"

Data available at: http://gvelarde.com/a/data.html.

It is worth noting that the survey did not provide options for people to choose between techniques used, functional applications, and fields of application. Although the word clouds obtained are based only on responses from 21 self-identified professors and post-docs in AI community channels, for the first question, respondents agree with WIPO statistics that deep learning is a relevant trend.

The answers to the survey's first question show that functional applications were more relevant than application fields. In contrast, to the second question, respondents focused on broader objectives. They gave more importance to ethics and data, humans, machines, the world, and a better understanding of how AI works and how to ensure good practices, mentioning: reproducibility, reliability, explainability, and generalization. Respondents also focused on specific application fields in medicine and healthcare, energy, climate, politics, transportation, market, business, and social issues.

The participants' shift in focus between current research trends and the AI research ideal could be explained. In some cases, research projects are funded based on proposals that take time to be accepted and must meet an objective that new challenges may have overshadowed. Besides, research may not always respond to ideals but to other interests, for example, commercial ones.

Real and Artificial Stupidity

We, humans, are the most intellectually developed species. We are also the most complex species and the one that can most efficiently cause not only its self-destruction, but that of other species. While there are accepted definitions of intelligence, it is difficult to find a definition of stupidity. Is stupidity the opposite of intelligence, or is it the lack of it?

In his 2016 book *The Neuroscience of Intelligence*, Richard Haier explained that "stupidity is not yet a category recognized by the National Institutes of Health", and so there is no institute that studies it, like intelligence is studied. Philosopher Avital Ronell stated in her 2002 book titled *Stupidity* that thinkers have confronted stupidity by admitting that it is "what is there" and cannot be identified, cannot be measured, and even the Gods could not fight it. At the same time, Ronell thinks that stupidity would be related to "the most dangerous" human forces and failures that deserve shared custody. Could it be then that intelligence and stupidity can coexist?

The artificial intelligence winters around the 1980s can be explained by the exaggerated enthusiasm of some researchers who celebrated excessively preliminary results and generated expectations they could not meet. Nils Nilsson explained that the enthusiasm of the media and investors faded as the expected results did not materialize; funding for artificial intelligence was reduced, and many researchers even stopped using the *AI* label to continue with their careers. In an article published in 1976, Drew McDermott questioned whether this field of research reached stupidity and called on the scientific community to be more prudent and responsible with their research and its implications. Then in a congress in 1984, he pronounced his concern about the very high expectation around intelligent algorithms. He reminded researchers to be disciplined and to educate the public so that another artificial intelligence winter does not occur.

For now, everything is going well, but some wonder if there could be another AI winter ahead. Indeed, many technical problems have been solved thanks to, first, the algorithms' development; second, the increased hardware capacity which is greater than a few years ago; and third, the vast availability of datasets to extract the new gold. However, what can be learned to make predictions and decisions is not only reflected in the data—which can be biased—but also in the subtle way artificial intelligence depends on the hypotheses of those who develop algorithms.

Researchers Nagpal, Singh, Singh, and Vatsa from the Indraprastha Institute of Information Technology in India published in 2019 an investigation to know whether deep learning networks can suffer from being biased, because in 2018 Amazon's facial recognition algorithms identified 40 percent of Black representatives of the USA Congress as potential criminals and also in 2015, Google applications for image recognition identified a Black couple as if they were gorillas.

Nagpal and her colleagues explained that cognitive-neurologists have long recognized the phenomenon of in-group bias, which makes it difficult for us to identify individuals of races or ages with whom we have little relationship or exposure, see illustration in Figure 1.9. The researchers wondered if the algorithms could suffer from the same effect, so they performed 36 experiments using four deep networks, sometimes pretrained and sometimes trained from scratch using over 10 million facial images. They found that deep learning algorithms exposed mostly to a particular group will encode specific characteristics of the group they were exposed to, and suffer, like humans, from the in-group bias effect when used to recognize images of groups to which they were little exposed during training.

Circle, what do you recognize?

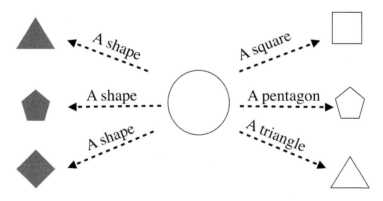

Figure 1.9: In-group bias.

Conference	Year	Female authors %	Female keynotes %	Female organizers %
NeurIPS	2019	10	43	52
IJCAI	2019	12	43	40
RecSys	2019	13	100	23
ICML	2019	14	67	38
ISMIR	2019	17	50	32
ECAI	2020	22	40	47
ACM FAccT	2020	42	67	50

Table 1.4: Percentage of female authors, keynotes and organizers in the following artificial intelligence conferences: Neural Information Processing Systems (NeurIPS), International Joint Conferences on Artificial Intelligence (IJCAI), ACM Recommender Systems conference (RecSys), International Conference on Machine Learning (ICML), International Society for Music Information Retrieval (ISMIR), European Conference on Artificial Intelligence (ECAI), ACM Conference on Fairness, Accountability, and Transparency (ACM FAccT). Based on data from divinAI.

It is believed that with greater diversity in research teams, problems of bias in automated systems could be avoided. Diversity is a topic of latent interest in artificial intelligence communities that join efforts to improve their diversity. In recent years, they established programs dedicated to promoting Women, Black, Latin American, Queer, Jewish, {Dis}Ability, Indigenous, and Muslim participants, as is the case of the diversity and inclusion initiative of the NeurIPS conference, formerly called NIPS, and which attracts thousands of researchers every year.

The percentage of female authors who publish in AI conferences is low. Still, since a few years, the organizing committees (which generally show better female participation at this level), have established a series of activities to promote greater female participation; for example, ensuring that the main scenario for keynotes is shared almost equally between binary genders. Yet, as seen in Table 1.4, the bias towards giving visibility to women may have to do with policies to promote this gender.

Besides that, there is little diversity in the origin of artificial intelligence inventions and publications. According to WIPO data, just over 60 percent of patents come from China and the USA, which share that percentage almost equally, followed by Japan with 17 percent. Likewise, China and the USA also share almost equally more than 40 percent of scientific publications on artificial intelligence. It follows the United Kingdom with six percent, India and Japan with five percent, Germany, France, Canada, Italy, and Spain with percentages lower than five percent, which indicates that there is little diversity regarding the origin of research.

The study of bias in artificial intelligence is a relatively new field. If in 2015 the scandal of the automatic labeling of Black people as gorillas occurred and in 2018, there was still a system biased towards recognizing a high percentage of Black representatives of the USA Congress as potential criminals, there likely exist various prejudicial "intelligent" systems operating in the market. By relying on the results of those "intelligent" algorithms, users will be making, unknowingly, biased decisions. Worst of all, "intelligent" automation will contribute to harm groups such as Women, Black, Latin American, and other minority groups, not only in the AI communities but also outside of them, through commercial or social service systems. Biased systems may contribute to worsening the gaps in education, employment, civil rights, or financial services, if no concrete steps are taken to address this problem.

Bias in intelligent systems can be seen as a problem of poor design, which will influence poor decisions by those who use those systems. In a more general view, poorly designed systems present the danger of causing accidents, that is, situations that occur unexpectedly and cause damage outside a laboratory. Considering safety in the design of intelligent systems is essential to avoid unwanted side effects. It should also allow systems to grow in a scalable, safe, and robust way in environments even for which they were not originally designed. In a 2016 article, Dario Amodei and his colleagues reviewed these considerations about problems in artificial intelligence safety.

Also, in recent years, ethics in artificial intelligence gained greater attention. Organizations such as the Association of Computer Machines, the Institute of Electrical and Electronics Engineers, the European Community, and the Organization for Economic Cooperation and Development presented artificial intelligence norms, principles, initiatives, and ethical guidelines.

Clever Hans Algorithms

In 1911, H. M. Johnson published an article in a journal of philosophy, psychology, and scientific methods, which described how a committee of 13 people, including two psychologists, two educators, two prominent zoologists, a veterinarian, and a circus director, gathered to evaluate the amazing skills of a horse. The horse belonged to a retired school teacher, who, over four years, taught his horse to solve exercises with the same method he used to teach in school. Hans, the horse, learned to communicate by indicating with his leg; he could add and multiply. For example, when asking the horse the result of a sum, Clever Hans moved his leg as many times as a result was. The horse also indicated the letters of the alphabet, and could distinguish colors, as well as musical harmonies. Hans the horse was so clever that he not only answered his instructor's questions but could also answer questions asked by strangers. Neither animal trainers nor scientists noticed any tricks performed by his instructor, who was, in fact, a respectable retired teacher from a German school, and who did not accept the large sums of money that people offered him for the horse. At that time, people considered Clever Hans had the intellectual faculties of a 12-year-old school child; and became very famous throughout Germany.

The evaluation committee took two days to perform some tests on Clever Hans. Some tests were performed in the presence of Mr. Wilhelm von Osten, his owner and instructor, and other tests in his absence. The evaluation committee concluded that there were indeed no "tricks" and the training method was based on the school method, but further investigation was required. Months later, the horse was studied again, this time with a smaller team of researchers. Based on further evidence, the researchers found that the horse responded to the questions with a 98 percent success rate if anyone present knew the answer, while only 8 percent of the responses were correct if the answers were unknown to those present.

So how did the horse know the answers? Did the horse have telepathic powers to read the minds of those who knew the answer? The researchers conducted additional tests, and finally realized that Clever Hans could actually read the body expression of people, who unwittingly and unconsciously made tiny movements of their head and neck, which the horse could read to give his answers, similar to many gambling people or mediums.

The truth is that Hans the horse was very clever! He could not add, multiply or recognize colors or harmonies, but he could read people's reactions. H. M. Johnson describes in his article that one pathetic outcome of the investigation was Mr. von Osten's painful attitude when he heard the scientific conclusions. He thought his integrity as a teacher, and his years of effort were being questioned, and he felt discredited by the scientific team. Mr. von Osten died months after having heard the research's conclusions about Clever Hans, without realizing his contribution to science.

An algorithm capable of giving correct results in a given problem, basing its answers on characteristics irrelevant to the problem in question, is like Clever Hans! For example, if the task was to classify music recordings, expecting an algorithm to learn to classify them based on their musical style, and you have a dataset with studio recordings of classical music and noisy recordings of folk music, and you find that the algorithm can accurately classify those songs, but fails when tested on another dataset that does not suffer from a noisy collection associated with one of the classes, then you could suspect, the algorithm was not learning to recognize music genre between classical and folk, but was using the recordings' noise of one class for its responses. Therefore it would be a Clever Hans-type algorithm, because it would not be able to recognize genre, which was the problem we wanted the algorithm to solve.

To prevent the creators of an algorithm suffering from syncope, or a very strong indisposition, similar to the one Mr. von Osten suffered when he discovered his horse did not solve the problems he thought Clever Hans did, algorithms must be evaluated with the same skepticism of the committee that evaluated Clever Hans. Even when researchers feel pressured by their clients to give a quick solution, or feel motivated by the current system, which promotes, not so much the quality of publications, but its *impact*, understood as the number of citations, not necessarily reflecting quality. Regarding this issue, in 2020, Yoshua Bengio wrote an article that calls on the scientific community to rethink the publication process, receiving several public comments. Also in 2016, Ludmila Kuncheva proposed a revolution in the review process, but we will not go into this issue.

Going back to the data topic, even in datasets where it would be assumed that there are no problems associated with the dataset collection, as in the previous example, where one class was systematically recorded with noise, and the other was not, there may still be characteristics in some datasets, that make an algorithm seem to solve a problem. But only through further testing, it is possible to determine the algorithm's merit.

In 2018, my colleagues and I published an article where we presented experiments to evaluate an ensemble of classifiers to recognize music style, either using audio or symbolic representations of music. We used *The Well-Tempered Clavier* by Johann Sebastian Bach, as it contains preludes and fugues that, although composed by the same composer, are stylistically different in terms of musical texture. We also used the string quartet movements by Joseph Haydn and Wolfgang Amadeus Mozart, since discriminating between those composers in the style of the compositions is a challenge not only for people but also for algorithms. To avoid presenting a Clever Hans algorithm or reporting "inflated" results, we performed several experiments.

Preludes are characterized by the harmonic elaboration of musical patterns, while fugues present and develop a theme in a successive and imitative manner like choir's voices, which in the initial section enter one by one and not simultaneously. This feature could be used to recognize musical texture between preludes and fugues, so we observed the effect of including and removing the initial section to avoid the problem of building a Clever Hans algorithm, and to avoid reporting results that are not generalizable. Indeed, when we included the initial section of preludes and fugues, the results were "inflated", faking better classification capability than when this section was removed.

Bob Sturm presented in 2014 a method to avoid falling into the problem of developing a Clever Hans algorithm, as well as Sebastian Lapuschkin and his colleagues in 2019, which method can help researchers understand what it is that algorithms learn when they solve a problem.

Taking Advantage of Limitations

In some tasks, such as chess, the general culture question game Jeopardy, or the strategy game of Go, artificial intelligence has demonstrated a better performance than the greatest human experts. The capacity of intelligent algorithms to solve a specific task is known as narrow Artificial Intelligence, narrow AI, or weak AI. In 2014, Ben Goertzel explained that in 2005, Ray Kurzweil would use the term narrow AI for algorithms that demonstrate "intelligent" behavior in a specific context. The concept of Artificial General Intelligence (AGI) would be popularized after 2007, with the publication of a book that carried that title. Artificial General Intelligence compares to human intelligence of general capacity, which can adapt and be flexible; it could even manage to solve problems in environments and situations their creators would not have contemplated. Artificial General Intelligence has not happened yet. Later on, we will examine when researchers believe this phenomenon could happen and what its impact would be.

Intelligence is difficult to understand. Rare cases have been reported of people with extraordinary abilities associated with an incredible memory, but at the same time, with severe mental disabilities. This phenomenon was described as the savant syndrome, and there are a few cases of people with it, such that some researchers dedicate several years to follow the savant they study. For example, Treffert studied for a long time the case of a savant who memorized over 6000 books, could give street by street directions on how to get from one city to another in the USA, and could calculate calendar dates. Yet, at the same time, he could not do tasks that would be simple for anyone, such as bathing or dressing himself.

Richard Haier considers artificial intelligence algorithms are more like a savant than a genius resembling Albert Einstein. Haier explains that human intelligence is usually estimated based on intelligence quotient (IQ) scores, with people considered to be more intelligent than others if their IQ score is higher. Furthermore, the various intelligence tests tend to be correlated, such that, for example, the result in a test that evaluates memory can predict how a person will do in other intelligence tests that evaluate spatial capacity or information processing speed and vice versa. While social or cultural factors can influence IQ tests, some researchers believe that the g-factor, also known as *general intelligence factor*, would be of a biological and genetic type. Savants are not considered intelligent, and their mental disability is clearly noted despite their extraordinary ability to memorize more than 6000 books, their splendid aptitude for drawing or sculpture, and even their fantastic ability for playing a musical instrument; in IQ tests, savants score low compared to ordinary people.

According to Haier, one of the most popular intelligence tests is the Raven's Progressive Matrices Test because it is an abstract reasoning test, independent of nationality, age, or language. Some researchers consider it capable of estimating general intelligence or

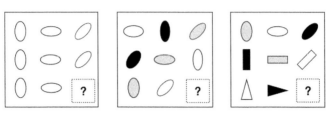

Level 1   Level 2   Level 3

Examples to be solved in the intelligence test. In each matrix, it should be indicated which figure corresponds to the box in the bottom-right corner ? *In level 1, the figures rotate by column. In level 2, the figures have certain orientation in the left diagonals (↘) and have a certain color in the right diagonals (↗). In level 3, the figures change by rows, rotate by columns, and take colors in the left diagonals (↘).*
The answers are: Level 1: ⬭ *, Level 2:* ⬬ *and Level 3:* ◁

Figure 1.10: Raven's Matrix Intelligence Test. Examples based on Mańdziuk and Żychowski (2019).

the *g*-factor. The test consists of deciphering patterns, as seen in Figure 1.10, which presents three levels of difficulty. The first level is the simplest, and the third level is the most complex, and at each level, you must indicate what should go in the box at the bottom right.

If Raven's intelligence test can estimate general intelligence, it would be convenient to evaluate whether the scores obtained by artificial intelligence algorithms are similar to those obtained by savants or by ordinary people. Mańdziuk and Żychowski from Warsaw University of Technology, Poland, developed a deep artificial neural network named DeepIQ, and proved that human IQ is higher than DeepIQ's score by 17 points at the simplest level. Still, at the second level of difficulty, DeepIQ manages to equal human performance. At the third level of complexity, the intelligent machine is superior to humans by 18 points, as shown in Figure 1.11.

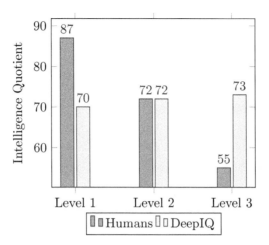

Figure 1.11: Raven matrix test results in three levels of difficulty obtained by humans and DeepIQ. Based on data from Mańdziuk and Żychowski (2019).

How to understand these results? Humans are far superior to intelligent machines in recognizing simple patterns, but the more complex the relationships between patterns, the lower our performance, while DeepIQ is immune to complexity. On the third level, we, humans, seem mentally handicapped in front of the machine and would not like to know our results on a fourth level of complexity. Could it be that these results explain what is understood as common sense or intuition, attributes that make us superior to intelligent algorithms, but at the same time, it shows that machines are really superior to us in finding complex patterns? I do not know. With these results, we could say that algorithms are getting closer and closer to reaching Artificial General Intelligence if Raven's test really estimated it, but we could also think that Raven's test lacks other features to be able to estimate the famous g-factor or general intelligence.

On the other hand, the fact that humans are better than intelligent algorithms at certain tasks and intelligent algorithms at others, is not a negative result, but rather suggests that we can use the difference in abilities to increase the overall capabilities. Even if in the future, improved versions of DeepIQ will demonstrate that the algorithms are superior to us at all levels in IQ tests.

IQ tests have been criticized on the grounds that they do not measure intelligence, but rather specific abilities valued by test designers. Also, it was observed that higher scores on intelligence tests are related to better socioeconomic status, which in itself possesses excellent benefits in all aspects of life. In environments where people attribute value to IQ tests, the most advantaged people are precisely the most benefited.

James Flynn, Emeritus Professor of the Department of Politics at the University of Otago, explained in his 2018 reflections that since the introduction of IQ tests between 1932 and 1976, test scores improved massively by 14 IQ points. Until 1987, massive increases were also seen in 14 countries. Now this phenomenon has been repeated in 34 nations and is known as the Flynn effect. As early as 2007, Flynn showed that gains in IQ did not mean that people a few decades ago were less intelligent than their children or grandchildren; or that there had been gains in g, general intelligence. The massive rise in IQ scores would be explained by the environment of an industrialized society that can take the "hypothetical seriously" and is faced with a more complex world, with more schooling, better nutrition, and with more people in occupations that demand cognitive abilities of abstraction, logic, and classification. Also, the research of psychologist David Marks, published in 2010, shows that IQ varies with the degree of literacy, and the current differences in IQ scores between races will disappear when people have the same opportunity to acquire similar literacy levels.

Are we not just more than seeds, each with its own genetic code but dependent on falling into good soil to grow and become lush trees? But also, adversity makes us stronger, more creative, and more resilient. Until the gap in IQs between advantaged and disadvantaged groups disappears, we will be further and further from achieving the performance and capacity that improved versions of intelligent machines will demonstrate when processing complex patterns. However, even if DeepIQ and other intelligent machines demonstrate their superiority in the most complex pattern recognition tests, it is necessary to consider that intelligence is not only related to the ability to recognize complex patterns. To survive, intelligent beings also possess other capabilities such as empathy, which is the ability to feel what others feel, understand them, and tune in to them.

A 2007 study by Herrmann, Call, Hernández-Lloreda, Hare, and Tomasello evaluated human infants of approximately two and a half years, chimpanzees and orangutans. On tests of physical domain concerning space, quantity, and causality, human babies and chimpanzees scored similarly and much better than orangutans, while on social tests of social learning, communication, and theory of mind—which is the ability to understand the thinking and feeling of others—human babies scored much better than chimpanzees and orangutans. Researchers observed that possibly a highly developed ability in humans is the ability to understand the causes of others' behavior as well as their mental states, for example, their intentions, motivations, or perceptions.

Also, Burkart, Schubiger, and van Schaik explained in a 2017 publication that general intelligence is not exclusive to humans, as studies with other animals show that rodents, primates, and birds exhibit general intelligence. However, they recognized that research in this area is still young and needs to mature. Researchers of the evolution of general intelligence explained that the results of studies done with humans and other animals generally did not include social cognition tests, but when included, the results were not conclusive. Therefore the authors considered it a priority of research to better understand the relationship between general intelligence and social-cognitive abilities.

From Perception to Creativity

Computational perception was one of the great challenges for intelligent machines, which failed miserably at tasks that are intuitive and fundamental to people, such as vision and hearing. Without the ability to see and recognize between the appearance of a deer or a tiger, our ancestors would not have survived for long, and likewise, without the ability to create, we would not be where we are.

Human societies have achieved significant development milestones thanks to creativity. Creative minds are celebrated in art, science, or business. There is a relationship between intelligence and perception, which are also related to creativity. Still, creativity is more difficult to evaluate than intelligence from any point of view, whether philosophical, neurological, or computational. Creativity is a capacity of intelligent entities to generate something new with its own value.

Philosophers Paul and Kaufman asked themselves: "What is creativity?" They considered that a person, a process, or a product, can be creative. Either an idea in someone's head or an observable, novel, and valuable object or situation, as Immanuel Kant had anticipated, for whom artistic genius produces not only original but exemplary works.

In the area of Computational Creativity, evaluating creativity is a major challenge for researchers despite the fact that there are several evaluation frameworks. In 1998, Margaret Boden, whose voice is influential in the debate on artificial intelligence and computational creativity, stated that creativity is a fundamental human characteristic as well as a challenge for artificial intelligence. She predicted that intelligent algorithms would find it easier to become creative than to be able to automate their evaluation. So far, Boden's observation is true in addition to being a logical observation, since evaluating creativity is complicated.

Creativity is not only observed in humans. In nature, there are many examples of intelligent entities that are creative. For instance,

Bronsema, Bokel, and van der Spoel studied the African termites' solution as a guide in air conditioning constructions without energy. Termites solved that problem in extreme climatic conditions without human architects or engineers, thanks to their intelligence and creative capacity. The researchers explained that termites cultivate a fungus for food which must be kept at 30°C (86°F) while the external temperatures of their architectural-engineering constructions are 0°C to 40°C (32°F to 104°F). Termite constructions did not exist in nature. Termites created those constructions, so we can say that termites are creative and their constructions are creative because they solve a problem masterfully.

How new must something be to be considered a creative product, and how do you measure its value? A work of art does not solve a problem like termite constructions. A piece of art has value because of its beauty or because it fulfills a function that makes sense to a group of people; that is, its value is given by cultural appreciation. Today, Vincent van Gogh's works are auctioned at exorbitant prices, while the artist lived in poverty and without recognition. The post-death recognition of van Gogh's work and genius is not an isolated irony, but rather a possibility latent in the arts, philosophy, science, or any cultural product. Personalities whose creations enjoyed little recognition in life and *post-mortem* celebration include painters such as El Greco, Paul Gauguin, Claude Monet, Frida Kahlo, writers Emily Dickinson and Franz Kafka, mathematician and astronomer Galileo Galilei, chemist Alice Ball, geneticist Gregor Johann Mendel, philosopher and historian Henry David Thoreau, inventor Hedy Lamarr, first programmer and precursor of artificial intelligence Ada Lovelace, among several others.

However, some creators achieve fame and fortune in life, as was the case of Salvador Dalí, who defined himself as "the embodiment of surrealism", considered that the difference between a madman and he, was that he was not crazy. He also considered himself as a mystic and used to talk of himself in the third person. In a 1971

Archivo Televisa News interview, Dalí said (in Spanish, translated as follows):

> Lately, I was questioned, what was the difference between a very good photograph, the best in the world, very real and naturally objective, and a Velázquez painting, which as you know is almost photographic, because there is no difference, there is no deformation. That is, if a photographic camera is placed in the same place as the painter's eye, well, the result is very identical, apparently. And then Dalí replied as always, in a very brilliant way, that the only difference between the best photograph in the world and a Velázquez painting, the only difference: exactly the difference of seven million dollars, because the wonderful portrait that the Metropolitan Museum of New York now has of Juan de Pareja had just been sold for seven million dollars. So the difference between painting and photography is precisely that a photograph is made by a completely mediocre mechanical eye made in Japan or in Cleveland or anywhere, and instead, a painting is made through a quasi-divine eye created by God.

To this statement, we could say that, in fact, a camera is also created by humans, and therefore is in itself a miracle of creation. Furthermore, a photograph can be taken by an artist. In 1948, Dalí himself posed for photographer Philippe Halsman, whose work was called "Dali Atomicus", in which it is possible to see Dalí suspended in midair painting; three cats jumping in the air as if they came out of a Dalí painting propelled by water, and also other objects floating in the air. Indeed, there is a difference between a masterpiece and other creations, and the value attributed to a masterpiece is not necessarily reflected in an economical amount.

Intelligent algorithms are subject to training, testing, and evaluation, despite the difficulty of recognizing whether something is creative or demonstrates creativity. Evaluating how good an algorithm is for classifying photographs of men and women is easier than evaluating whether an image generated by an artificial intelligence algorithm can be considered a product of creativity, a true work of art, or not. Algorithms capable of creating are called generative models. Many times researchers declare their algorithm a success of computational creativity if it reaches a low measure (e.g., cross-entropy), other times if a group of people accepts the generated product, or if it deceives people who cannot tell whether it seems to have been created by a machine or not.

In 2007, Graeme Ritchie proposed some parameters to evaluate computational creativity, considering aspects such as novelty, typicality, and quality. The novelty criterion measures how similar or different a product is to other products in its category. Typicality measures how exemplary a product is to other products in its class, and quality measures how valuable a product is to other products of a similar nature.

The applications of generative algorithms are as varied as are how they are evaluated. For example, algorithms have been developed to automatically write "fluid and coherent" Wikipedia articles that contain "objective and relevant" information. Images generated by artificial intelligence algorithms were auctioned in art galleries for almost USD 100 000. Most people do not discriminate better than random guessing if a piece of music was composed by Johann Sebastian Bach or by an intelligent algorithm. And, it is difficult to distinguish between face images of real people and images generated by algorithms.

Creativity's philosophical study is taking more interest than ever before, and artificial intelligence is becoming more and more creative. Some think that creativity is a uniquely human feature and try to take comfort in the fact that it would be "the difference" between humans, other animals, and intelligent machines. In contrast, others question the creative abilities of the most creative humans compared to those of machines. But, while the discussion about what makes something "truly" creative, artificial intelligence algorithms are becoming more surprising day by day.

According to Margaret Boden, true or real creativity has to do with intentionality, autonomy, self-evaluation, emotion, and consciousness, but indeed, some of these concepts are difficult to define.

Computational creativity has always been a topic of interest in artificial intelligence. From the fantasy described by Aristotle of lyres playing their own chords, to the complex and creative scopes of analytical machines envisaged by Ada Lovelace, and today's commercial success of artificial intelligence's creations. From algorithms that challenge experts' strategic and creative capacity in chess or Go, to considerations of researchers Colton and Wiggins on computational creativity as "the final frontier" for artificial intelligence.

The Panic and the Love of Automation

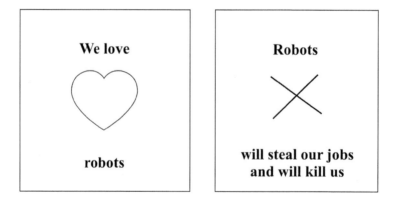

Figure 1.12: Acceptance and fear of intelligent automation.

When Jacquard looms appeared in 1801, English artisans and weavers who over time had become skilled in their craft as weavers panicked, organized and called themselves Luddites. Lord George Gordon Bayron, a famous poet of his time and father of Ada

Lovelace, supported artisans in their struggle against the introduction of mechanical textile machines. These machines threatened to replace trained artisans with people without any special skill, who could easily operate the new mechanical looms of high textile production. In contrast, mathematician Hannah Fry suggests that Ada Lovelace would be fascinated by Jacquard looms and their binary instruction technology that could mechanically create complex textile patterns. By 1811, Luddites united to destroy the expensive textile machines and even to burn down entire factories. As the vandalism continued, the English government introduced severe measures such as the death penalty for those who destroyed the mechanical textile machines, so that by 1813, Luddites' protests died out.

Today, the perception and acceptance of robots in society vary according to media exposure, religious beliefs, social dynamics, and cultural norms, as a 2014 study shows. East Asian countries such as Japan, where government, academia, and industry promote the acceptance of robots, robots are highly appreciated, even as "invaluable friends". In contrast, in the USA, intelligent automation is often negatively associated with unemployment, and with images reminiscent of a half-organic and murderous robot like the one in the film *The Terminator*.

In 2014, H. R. Lee and Šabanović published a study based on interviews of participants from South Korea, considered as a technologically developed East Asian country; Turkey, a rapidly growing Middle Eastern economy; and the USA, a developed Western country. The results showed that while all participants believed that media portrays robots as useful and intelligent, in the USA robots are associated with more negative and even frightening perceptions than in South Korea and Turkey, where robots are associated with more positive and friendly perceptions, and would be more accepted in those countries in social interaction scenarios than in the USA.

Those concerned about the harmful effects of intelligent automation, wonder which occupations will be stolen by robots and in which sectors. Others, instead, expect that robotization will drive global economic growth thanks to intelligent automation, increased worker productivity with better work tools, and the emergence of innovations with new revenue streams, see Figure 1.12.

In 2018 there were several predictions about how intelligent automation will look like for workers. Agrawal, Gans, and Goldfarb, economists from the University of Toronto, argue that intelligent machines will make predictions: cheaper, faster, and better, and complementary tasks such as data collection and decision-making will increase in value and remain as human tasks.

However, a 2018 English study by PricewaterhouseCooper's consultants estimates that by 2030, up to 30 percent of jobs across all sectors will be automated, but some sectors will be more affected than others. For example, 50 percent of jobs will be automated in the transportation sector, 30 percent in financial services, and 20 percent in healthcare. Besides, smart automation will replace more than 40 percent of workers with low education, 35 percent of workers with medium education, and 10 percent of those with high education. Other economists estimate that in the next few years, only 16 percent of professionals in Peru would be adequately prepared to face the current technological revolution. In Mexico, strategists estimate that 19 percent of jobs will be affected by intelligent automation in the manufacturing, construction, retail and wholesale, agriculture, hospitality, and foodservice sectors.

On the other hand, in a debate organized by the World Economic Forum in 2018, various representatives from Asia showed a very positive view about intelligent automation. Asian experts recognized that it allows a large number of people to offer their products and services to a large number of buyers. The new business models are growing in record time, and instead of talking about a force of "creative destruction", they consider that the current technological revolution is an "organizing force" that allows people to connect and work, also allowing governments to apply regulations and taxes to those who previously worked in the informal sector.

Until 2016, the most robotized countries had the lowest unemployment rates, see Figure 1.13. As reported by the International Federation of Robotics, countries with the highest number of robots operating per 10 000 employees in the manufacturing sector in 2016 were South Korea, Singapore, Germany, and Japan, which according to the Organization for Economic Cooperation and Development (OECD), had the lowest unemployment rates in 2016. Although unemployment is a complex topic that cannot be attributed to the lack of robotization only, intelligent automation will take over several trades. Still, new jobs will appear to accompany the innovations of artificial intelligence.

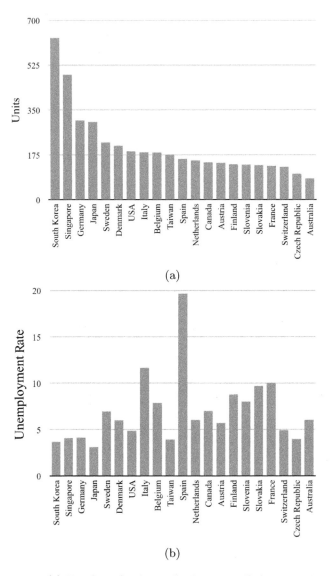

Figure 1.13: (a) Number of industrial robots installed per 10 000 employees in the manufacturing sector in 2016. (b) Unemployment rate in 2016. Based on data from the International Federation of Robotics (2017) and OECD (2020).

Chapter 2

Innovate, Adapt, or Go Extinct

People have a high capacity to adapt to the environment and exceptional environments allow the development of exceptional talents. To what extent can we overcome the shortcomings of an environment or take advantage of its goodness, and what strategy could we follow to not only survive an Artificial Cataclysm but to prosper thanks to it? What role will entrepreneurs, scientists, companies, or governments play during the current technological revolution, and how can they catch up or prepare themselves to enjoy the benefits of cutting-edge technologies in the near future?

Innovation and the ability to adapt quickly to the environment will be essential in the coming years. Innovation is synonymous with improvement, marketing, or with implementing something improved at the cost of possibly causing a previous solution to vanish. In the 1934 theory of economist Joseph Schumpeter, entrepreneurs are the "heroes and captains" of the market and those who create "new industrial empires" in a process that includes daring to innovate, selling new products, and introducing technological improvements opening new paths; furthermore, for Schumpeter, innovation is essential and more relevant than the struggle between competitors to lower prices, because innovation promotes economic development, which in turn explains "capital accumulation" in a constant process of "creative destruction".

Is the Artificial Cataclysm an opportunity or a threat? In the next 15 years, the technological revolution could represent a threat for some, while it could be an opportunity for accelerated growth for others. The small ones cannot follow the same strategy the giants will follow and vice versa. Who will depend more on whom to survive and how to collaborate to ensure a win-win coexistence will be an art.

Expectations about the impact of this technological revolution in the medium to long term are varied, and in some cases, diametrically opposed. In recent years, new technological business models have painted a flag of optimism in some territories, especially in Asia, while in Western countries, the positions are opposite: between optimists and pessimists, such that there is concern about whether intelligent automation will put many people out of work, or whether inequality gaps will widen further, versus positions describing that in past technological revolutions labor productivity increased and more jobs were created than lost. Some countries seem to be watching the landscape closely, as attentive as in a tennis match, while others have already taken action because they are confident that an exceptional opportunity to grow, develop, or achieve leadership exists.

The current scenario is fast-paced, with an explosion of scientific discoveries and technological innovations. A new *gold* has made possible the sudden creation of large companies, and it will be essential to make good decisions of great agility, flexibility, and vision, as the expected changes are so profound that some empires will disappear, while others will emerge, even a "New World Order" is expected.

Fortunes of Yesterday and Today

The chance discovery of a treasure would be anyone's fantastic dream. For outsiders looking to become millionaires overnight, Potosí was the ideal destination. At the foot of Cerro Rico, Potosí is home to a Bolivian mining town where the world's most incredible fortunes were made in colonial times. At about 4 000 meters above sea level, some actually lived closer to heaven with all the wealth that others extracted from the fiery pits several meters below the ground, and under the tutelage of *"El Tío"* (The Uncle), the Andean deity who rules in the depths, protects, rewards, and punishes, and to whom offerings are still made at the entrance to the mines.

In the centuries of colonial exploitation of Cerro Rico, so much silver would have been extracted from it, that historians suggested that a "silver bridge" could have been built from Cerro Rico in Potosí to the Royal Palace in Madrid, a straight line distance of about 9 300 kilometers (approximately 5 780 miles). The popular saying adds that another bridge could also have been built parallel to it with the bones of miners who died working in the mines. According to Eduardo Galeano's research, it is possible to imagine the wealth's magnitude that Cerro Rico de Potosí had. Only from 1503 to 1660, the accounting showed that "185 thousand kilos of gold and 16 million kilos of silver" arrived in Seville—not counting smuggling, piracy, and the ships that sank on the way to their destination. These precious metals were extracted by miners who worked practically for free and in inhumane conditions despite the existence of official indigenous labor laws; in the mines of Potosí, the foremen used the saying that still resonates: *"La ley se acata, pero no se cumple"* (The law is obeyed, but it is not enforced).

The extraction of gold, silver, and other metals would continue under colonial conditions until the end of Bolivia's war of independence in 1825. Potosí became one of the richest cities in the world at the beginning of the seventeenth century and, according to Galeano, the high societies enjoyed an exuberant wealth, where silver was used even in horseshoes or as decoration in houses also ornamented with gold. You could find all kinds of luxuries brought from all over the world, including diamonds, rubies, or pearls.

Popular stories tell that the rich people of Potosí used to keep their treasures—gold coins, silver, precious stones, and other jewels—in wooden or leather chests, which would be hidden in the thick adobe walls of Potosí's colonial houses, under the floor, or even inside grand pianos. These types of treasure chests are called *tapados*. Some tapados survived their owners and remained hidden for years and even centuries. Every Potosino knows that there is still a remote possibility of finding a tapado full of silver, gold, and jewels in the walls or foundations of a colonial house. Every other Potosino knows the story of an acquaintance who became rich, or disappeared from the map, after finding a tapado. Although gold no longer emanates from Cerro Rico de Potosí, the city is still a mining town but a low-income one.

Today the new gold is data that data miners extract from the great activity over digital platforms, or from sensors and machines, in industrial plants or hospitals. With this new gold, empires like Google, Baidu, WeChat, Facebook, Amazon, or Twitter were suddenly built and empires like IBM, Toyota, Bosch, Siemens, Philips, or Samsung are being empowered. Tech companies employ the brightest scientists to forge the smart path of their businesses.

Although the largest populations on the planet are in China, India, and the USA, according to Statista, the number of Facebook's active users almost doubles the Chinese population, WeChat's active users are close to the Indian population, and Twitter's active users almost equal the population of the USA, as seen in Table 2.1. Observe the large growth of Internet platforms since 2015. Besides, Alibaba's number of active users via mobile phone was about 743 million in the second quarter of 2020. And Amazon's Prime paying users reached 150 million in 2019.

Country or platform	Population in millions in 2015	Population in millions in 2020
Facebook	1 400	2 701
China	1 406	1 439
India	1 310	1 380
WeChat	697	1 206
USA	321	331
Twitter	307	330 (2019)

Table 2.1: The largest populations on the planet in millions, considering the three most populated countries and the number of active users of three of the largest Internet platforms. Based on data from Worldometers (2020); Statista (2020c, 2020b, 2019).

From Bedrooms and Garages to Empires

A documentary by Klein and Pradinaud describes how a brilliant student of computer science and psychology at Harvard University accessed digitally, and without authorization, university records from his dorm room in a student residence. With the students' photographs, he created, in collaboration with another classmate, a website called FaceMash, which launched in 2003. FaceMash caused curiosity and commotion because it invited the user to choose: "Who's Hotter", between pairs of photos from the students on campus. As seen in the documentary, the hacker did not consider it a bad action to have demonstrated the weaknesses of the university's security system. The deans, although upset, did not expel him but granted him a final warning for his action, and after the sentence was given, the student celebrated his feat with champagne. That student was Mark Zuckerberg, who in a few years became the president of Facebook, and, in 2020, became the third richest tycoon after Bill Gates and Jeff Bezos, as reported by Bloomberg.

Two other students enrolled in a computer science doctoral program at Stanford University, turned their university project into a multimillion-dollar business just in five years after founding Google in 1998. Larry Page and Sergey Brin were studying how to make

Internet search results more relevant and faster. To work on their project, they rented the garage of entrepreneur, economist, historian, and mother of five Susan Wojcicki, who would later become part of the Google team and suggested the purchase of YouTube, becoming the president of this company since 2014. Both, Facebook and Google, were founded and flourished in the famous Silicon Valley of the USA, the mythical place of great cultural diversity where startups, investors, and universities work tirelessly to export innovations to the world.

On the other side of the globe, in China, Jack Ma, an English teacher with a vision, described in a 2015 documentary by Erisman how he decided to bring 17 friends together in an apartment and convince them they were as capable of creating a digital empire as the people of Silicon Valley. Ma also sought to engage his government in conversation, because he was aware he needed their support so that China could enable world-class digital platforms. Today Alibaba, the company that Jack Ma founded in 1999, is one of the largest e-commerce companies in the world.

Kai-Fu Lee, a businessman and computer scientist who currently runs a venture capital fund in an area of Beijing called Zhongguancun—the Chinese analog of Silicon Valley—predicted that China will be the next AI-powered empire: China's government developed an ambitious strategy to exploit it, China's large digitally connected population allows for the collection of vast amounts of data; Chinese students are consumed by studying artificial intelligence, and the Chinese entrepreneurship ecosystem is, for Lee, much more fierce than that of Silicon Valley. Lee also explained that until a few years ago, Chinese technologists were known for replicating successful USA platforms and business models, and people used to talk about China's Google (Baidu), China's Facebook (Xiaonei), or China's Twitter (Fanfou), but today, Chinese companies are just like or even more innovative than USA companies. For example, WeChat had 300 million users by the second year of its launch in 2013, and more than a billion in 2018. WeChat users can send messages, make a doctor's appointment and pay for it, fill out tax forms, buy airline tickets, hold a video conference, and more, all in one app.

Currently, the USA and China are recognized as two powerhouses of artificial intelligence, as they have the highest number of artificial intelligence patent filings in the offices of the World Intellectual Property Organization.

From Empires to Ashes

Once an innovation occurred, previous versions are often destined to disappear. For example, in the music industry, cassette displaced vinyl record, compact disc displaced cassette, and streaming services displaced the former. A company clinging to continue producing compact discs without considering moving its business model to streaming services, could be bringing about its ruin. Many of the readers of this book will have never heard of Myspace, a social network, AltaVista, a search engine, or Kodak's photographic rolls, which resemble the ashes of a large bonfire.

Kodak is a classic case study in business because after being an international monster that captured the happy moments of several generations with its cameras, films, and developing machines, is barely even a shadow of its former self. Would Kodak have dug its own grave when it invented the digital camera? Or would it have ignored that Fujifilm knew how to make roll film faster and cheaper? John Bessant claims that above all, its leaders would not have reacted on time coordinating internal and external competitions to compete in a sustained manner.

Another example is the AltaVista search engine. When Larry Page and Sergey Brin from Google came on the scene, there were already seven Internet search engines, and one of the most popular was AltaVista. Before Google's search engine appeared, one could go to an Internet cafe, sit down in front of a computer connected to the World Wide Web, type something into the browser search bar, order a coffee, and drink it little by little until search engines gave you an answer that amazed you with its novelty, even if the results were not that relevant. AltaVista is practically history today.

For Nikhil Dandekar, the innovations by which Google surpassed the other search engines would be that Google was about 5 to 10 times faster than the rest of search engines, and its indexing returned many more relevant results. Google showed these two features at the top of the page to demonstrate its superiority. Their interface was clean and minimalist. Results were shown together with snippets that helped decide which results were more relevant to the user, without the need to visit each link. Besides, in their early years, they focused only on doing one thing well—searches—and after becoming the leading search engine, they expanded their horizon to data-center technology and monetization strategies.

At some point, the social network Myspace had more visits than Google and was later eclipsed by Facebook, which was then a startup founded by five people, including Mark Zuckerberg. According to Gil Press's 2018 analysis, Myspace did not restrict membership to its users and allowed them to configure their pages, losing its identity. In contrast, Facebook maintained a corporate format for user pages and controlled their growth with a robust infrastructure while avoiding the technical problems of competitors. Its reputation and innovation culture attracted experienced engineers to develop innovative applications. Furthermore, Press believes Facebook knows how to use the concept of social approval, and Mark Zuckerberg is the master of public relations. For example, in 2009, Zuckerberg asked its users to vote and "decide" on the platform's terms and conditions, like true Facebook "citizens". In addition to Press's points, I think the company kept an eye on dangerous platforms. Possibly, if Facebook had not acquired Instagram and WhatsApp in due time, it would be history.

Can an organization sense when its time will come, or can it recognize within the crowd the David who will overthrow it? The king of the legend would not have lost his kingdom if he had recognized the magnitude of the creator's request for his invention. The

1	2	4	8	16	32	64	128
256	512	1024	2028	4096	8192	16384	32768
65536	131K	256K	524K	1M	2M	4M	8M
16M	33M	67M	134M	268M	536M	1G	2G
4G	8G	17G	34G	68G	137G	274G	549G
1T	2T	4T	8T	17T	35T	70T	140T
281T	256T	1P	2P	4P	9P	18P	36P
72P	144P	288P	556P	1E	2E	4E	9E

Figure 2.1: Exponential growth, number of rice grains ordered by the inventor of chess, from 1 to 9 Exa (E) grains. Based on work by Andy0101 (2010).

inventor of chess made a simple request for his creation: for the first chessboard's square, a grain of rice, the next square twice as much as for the previous one, and so on, until the last chessboard's square, which has in total 64 squares, see Figure 2.1. The king thought it was an insignificant request for such an entertaining game, and granted the inventor his request, not realizing that what he was asking for, was a quantity of rice that followed exponential growth, such that the king could not pay him even with all his wealth. According to calculations presented in Wikipedia, the amount of rice the inventor of chess should have received is equivalent to a thousand times the rice production of the planet in 2011.

Strategies' Pillars

As explained in the first chapter, milestones marked by artificial intelligence have been happening increasingly in recent years. Two news items that seem to have had a profound impact were AlphaGo's victory over Lee Sedol in March 2016 and Vladimir Putin's comment in September 2017, about the relationship between artificial intelligence mastery and absolute dominance. That same year, artificial intelligence strategies began to appear everywhere worldwide.

Strategies are used to achieve a goal. Many countries have chosen to develop and implement their strategies since 2017, see Table 2.2. Additionally, other measures are developed to promote technology adoption, see Table 2.3. Some governments created an institution dedicated to directing and coordinating their artificial intelligence strategies and plans. For example, China, which is currently one of the leading producers and exporters of artificial intelligence researchers, has a new office within its Ministry of Science and Technology to execute its plan and strategy; the United Arab Emirates has a new Ministry of State for Artificial Intelligence to run its strategy; and Finland has a new group to implement its Programme as part of the Ministry of Economic Affairs.

I found recurrent and common themes among the artificial intelligence strategies in Table 2.2 which I present as the Fundamental pillars in Figure 2.2. The fundamental pillars can be addressed in parallel, and these focus on an education that prepares future experts, dynamic entrepreneurs and innovators; an adequate physical and digital infrastructure; good relations with experts, investors and other neighbors; agile and promotional regulation; and, excellent coordination. It is essential to consider that each strategy has its own specific priorities and pillars, and there is no one-size-fits-all solution for all countries: each one is at different levels of development and

Country	Strategy or Plan	Budget
Canada	Pan-Canadian AI Strategy	$125 million
Japan	AI Technology Strategy	No information
Singapore	National AI Strategy	S$150 million
China	New Generation AI Development Plan	No information
United Arab Emirates	National strategy for AI	No information
Finland	AI Programme	No information
Taiwan	Taiwan AI Action Plan	NT$40 billion
Denmark	Toward a Digital Growth Strategy - MADE	DKK 1 000 million
France	AI for Humanity	Euro 1 500 million
United Kingdom	Sector Deal for AI	Pound 300 million
Australia	Digital economy strategy	$29.9 million
United States of America	The American AI Initiative	No information
South Korea	Mid-to Long-Term Master Plan in Preparation for the Intelligent Information Society	Won 2.2 trillion
Sweden	National Approach for AI	No information
India	National Strategy for AI #AIforAll	No information
Mexico	Toward an AI Strategy in Mexico & Harnessing the AI Revolution	No information
Germany	AI Strategy	Euro 3 000 million
Spain	National AI Strategy	No information
Lithuania	AI Strategy	No information
Russia	National AI strategy	No information

Table 2.2: National AI Strategies/Plans by country in order of publication from 2017 to March 2020 (Velarde, 2020a). Reprint courtesy of Gissel Velarde.

Country	Type of action
Argentina	National AI plan in development
Austria	National AI strategy in development
Brazil	Creation of 8 AI laboratories and adoption of OECD AI principles
Colombia	Establishment of the Center for the fourth industrial revolution
Chile	AI policy under development
Estonia	Legal framework for the use of AI in development
Italy	Establishment of a specialized AI team
Ireland	Development of the AI Master's Program
Kenya	Creation of a team specialized in blockchain and AI
Malaysia	National AI framework under development
New Zealand	Creation of the platform AI Forum
Uruguay	National AI strategy in development
Tunisia	National AI strategy in development

Table 2.3: Government measures to promote artificial intelligence by country in alphabetical order, until March 2020 (Velarde, 2020a). Reprint courtesy of Gissel Velarde.

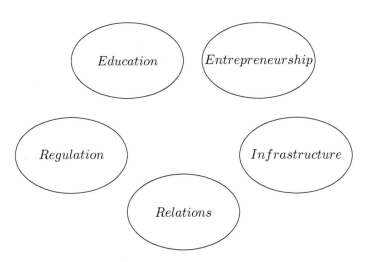

Figure 2.2: Fundamental pillars of artificial intelligence strategies.

has different characteristics. However, the pillars in Figure 2.2 can be considered as essential and can be applied at different levels and scenarios, from a one-person startup, to a large company, and from one country to a set of countries.

In the surreal world we live in, some people, groups, organizations, and countries seem to be better equipped to cope with the coming years' technological shock. Still, they all have strengths and challenges, and the option to collaborate or destroy each other. Whether the impact of intelligent automation will be positive or negative for society is one of the most debated issues; some trades will cease to be practiced, and although new ones will appear, inequality gaps could widen.

In 2017, Carl Benedikt Frey and Michael Osborne, both from Oxford University, presented probability calculations to estimate, within a list of 702 occupations, which are at risk of being automated, or in other words, to be deleted from the list of jobs and occupations to be carried out by people. But without the need to see the table of jobs and their probabilities of extinction, the simplest thing to do is to consider that such research found that low-wage occupations which require little training or education are the most susceptible to disappear. In any case, it is crucial to consider that in several countries, instead of having tables of occupations, they have some multicraft people, or people that work on tasks extinct in high-income countries. For example, not even automatic washing machines have completely replaced Bolivian laundresses, who, as reported by Opinión in 2018, wash about five to six dozen garments daily to support their families.

Investing in education in these bizarre times is the wisest thing to do. Some countries have a higher proportion of entrepreneurs than others, either out of necessity, by entrepreneurial opportunity recognition, or by a combination of both. Entrepreneurs, ideally well-educated, will need a secure system that supports them and at the same time benefits from their innovations.

A 2018 survey by the World Economic Forum, asked more than a thousand participants whether their education system prepared them for their current job. Only one-third of respondents believed it had, while two-thirds thought it had not. Indeed, for the majority to think that education systems do not prepare workers may indicate a lack of communication between education and labor sectors, a radical shift in industry needs, or simply that working life is a constant learning journey.

Conferences can be used as a mechanism for broad knowledge transfer between different sectors, bringing to light challenges and solutions from various perspectives. Alternatively, the very fashionable hackathons could help create links between universities and businesses if fair conditions are promoted. In 2017, Zukin and Papadantonakis criticized hackathons for being events where young people work furiously for "soda and pizza" (precarious working conditions), while arguably, those who benefit the most are the organizers.

Another essential element in today's strategies is to have a robust infrastructure to enable digitization so that further innovations take place. Digitization can alleviate several pains, such as corruption or bureaucracy, and can reduce many costs associated with the problems mentioned above. However, digitalization also comes hand in hand with other problems, such as cybercrime, loss of privacy, or monopoly.

For example, Denmark is a digitized country, where paper documents are almost no longer used, digital signatures are accepted, banks charge extra costs for transactions in their physical offices: Internet banking saves them on rent and staff costs, and since most people no longer use coins and bills, the few beggars on the streets, instead of having a can to collect alms, put up a sign with their phone number to receive a mobile payment transfer, and cybercrime is one of the options for criminals. The strategy of technological progress followed by Denmark may not be directly replicable in a country that has not yet digitized. Still, in several cases, it may be an opportunity to solve more than one problem in one go and in leaps and bounds avoiding previous stages of development. Stones in the shoe and a small pocket are an excellent combination to find spectacular solutions.

Those who are too fragile or too rigid will find it difficult to adopt or give continuity to their strategy during this industrial revolution. Technology can advance faster than regulation. Therefore, agility in decision-making is vital so as not to slow down or fail to take advantage of technological innovations. Still, many times innovations can have such profound impacts that no one is able to estimate. Innovators will need to speak the same language as regulators to understand each other and work together to promote and adopt new technologies. Strategists consider observing policies and processes that may impair or delay executing their strategies, for example, on issues related to startups, training, or experts and investors attraction. It would be a bad move to invest in developing a strategy and not have the power to execute it quickly.

Isolation does not seem to be an option, and building good relationships with experts and investors of neighboring or distant countries is a good idea if there is a genuine interest in collaborating and not leaving anyone behind. Excellent coordination will be essential and will surely be one of the most delicate tasks.

Strengths of the Small and the Large

There are general elements that everyone can use to a greater or lesser extent during this industrial revolution. For example, the energy of dynamic young people, the characteristics of businesses based on intelligent technologies, the current technological environment, and the intention of some people to create a better world by taking advantage of all of the above. Indeed, the widespread accessibility of information and intelligent technologies present a great transformative opportunity.

For economists along the lines of Schumpeter to Piketty, entrepreneurship and innovation are the engines of economic growth, and governments must protect entrepreneurs because of the value they represent. In the Global Entrepreneurship Report 2015, Querejazú, Castellón, and Córdova studied 70 economies and observed two relevant groups. The most dynamic group corresponds to early-stage business entrepreneurs (aged 25 to 34). The second group corresponds to established businesses owners (aged 35 to 44). The 10 countries with the highest proportion of their adult population in entrepreneurship were Cameroon, Uganda, Botswana, Ecuador, Peru, Bolivia, Chile, Thailand, Burkina Faso, and Angola. The top seven countries with the highest rates of opportunity-driven entrepreneurs were Denmark, Taiwan, Norway, Trinidad and Tobago, Luxembourg, Singapore, and Sweden. Young populations, combined with their entrepreneurial spirit and government efforts to provide proper training and technological tools, will be essential during today's technological revolution.

In digital and artificial intelligence-driven businesses, one of the most important investments is in the know-how or knowledge of your team, which is usually one of the magic ingredients of success. Compared to investments in other industrial sectors, what entrepreneurs will initially spend on equipment or infrastructure is low. With luck, a startup can grow into a rolling snowball.

Experts speak very optimistically about the scope of this industrial revolution in a debate on Asia's future published in 2018 by the World Economic Forum. Shailendra Singh from Singapore's Sequoia Capital said that because a large proportion of people can connect to the Internet with accessible smartphones, new businesses are growing in record time and in leaps and bounds. While some time ago it used to take startups about 10 years to become large companies, luckily today, it can take them between one to three years. Therefore, not only companies benefit, but also users, who intuitively learn to master mobile devices, access platforms, download applications, offer services and leave the informal sector to enter the formal one, allowing the enforcement of regulations and taxes, as explained by Thomas Lembong from Indonesia Investment Coordinating Board, who is convinced that the current technological revolution is more of an "organizing force" than a force of "creative destruction".

In countries where wages are lower than in others, the cost of producing a product or service will be in line with what workers earn because the most important expenses will be on their talent, that is, their people. Eduardo Galeano would talk about "cheap arms". Nowadays cheap arms are not needed, but rather prepared people with a certain level of education. We could talk about "efficient brains" capable of keeping operating costs down. Data miners can extract gold from data. Data is provided by users. Some entrepreneurs realize that it is "fair and necessary" to compensate for their extreme luck and success with a Universal Basic Income that could broadly benefit society on the planet.

CNBC featured a report in 2017, in which, dressed in an elegant suit and almost in a campaign tone, Mark Zuckerberg said, standing on a podium at Harvard, his alma mater and university that forgave him the FaceMash scandal, that "we should explore ideas like Universal Basic Income" and need a society that provides lifelong education, instead of a society that stigmatizes those who make mistakes, because "we are all going to make mistakes", Zuckerberg stressed.

Artificial intelligence creates fascination due to its versatility. It is a powerful tool that can be used for any purpose. Recently, some organizations are promoting its use as a sword of good. Since 2017, the United Nations annually organizes an event to promote its 17 Sustainable Development Goals. Other relatively new organizations announce the use of artificial intelligence in projects advancing humanity's well-being, for example PartnershipAI, OpenAI, and MILA.

Being Alert to Challenges

Being like a fly would involve being extremely alert, being able to quickly recognize changes and react to them, also knowing that you can be crushed at any moment. The challenges during this technological revolution are related to international competition, talent generation and retention, experts and investors attraction, and regulations.

There are consolidated ecosystems in science, technology, and innovation that are magnets for talent. Some countries' strategies explicitly contemplate the desire to become leaders in artificial intelligence, and some are already reaping the first fruits in this field. They traditionally host renowned scientists and successful entrepreneurs in their universities and companies. Additionally, they possess sufficient economic resources, and a state-of-the-art digital infrastructure.

During this technological revolution, human talent will be the most critical resource for businesses. The challenge will not only be to build capacity in emerging technologies quickly but create an attractive environment to retain that talent. The demand for professionals specialized in artificial intelligence and other essential skills in the coming years will be high. For astute and adventurous young people, it will not be a constraint to move quickly from one place to another where the salaries are more attractive, or the environment is more exciting, or the possibilities for self-development are better.

Until a few years ago, renowned researchers worked in academia because the horizons of excellence between academia and industry conflicted: high-impact scientists are recognized for their publications, and industry jealously protected their intellectual property. However, technology companies understood that it would be impossible to attract the brightest brains if they were not allowed scientific dissemination and found mechanisms to manage it with the acquisition of patents and commercial exploitation of results generated by their scientists. Today it is just as prestigious to be a professor at the best universities in the world as it is to be a research director at a company like Facebook, WeChat, Xiaonei, Google, Baidu, Yandex, Naver, Alibaba or Amazon. Moreover, the current conditions under which technology companies operate are sometimes more interesting for artificial intelligence scientists because of the large amount of data they have to carry out their research and because of the resources at their disposal to fulfill their whims. For example, Yann LeCun is a Vice President and Chief AI Scientist on Facebook, Corinna Cortes is a Vice President in Google Research, and Kai-Fu Lee is the CEO of Sinovation Ventures, formerly president of Google China. Companies are interested in the vision of their prestigious scientists to guide them to destinations of excellence.

Countries without state-of-the-art ecosystems, or midsized organizations, will find it difficult to compete with salaries and benefits of vibrant ecosystems, so they will need to understand other motivations for attracting talent, such as ideological motivations, responsibility and opportunities for career development, visibility, work–life balance, cultural environment, climate, and optimal conditions for relocation. At the same time, digital technologies have shortened the distances, and the coronavirus (COVID-19) pandemic showed us the need, benefits, and nuances of working online. In many cases, it is possible to collaborate with experts efficiently and inexpensively through video conferencing.

One strategy to attract talent and promote innovation is to set up startup programs, which are very fashionable. A startup would be different from another enterprise because of its technological approach that can grow quickly and widely, not only in a domestic market but also internationally. The dream of every startup is quasi to experience exponential growth. If the business model is linked to the number of users and their information traffic, server costs can kill the business, and therefore investors' support is necessary for this type of business model to stay afloat.

What do investors look for when considering putting their money in low-income countries? That is the question posed by Lamech and

Saeed of the World Bank, who published in 2003 a study in the energy sector. The study's surveys revealed that investors expect respect for their rights and political stability to ensure a return on their investment without interference from arbitrary governments that do not uphold proper contract terms but rather improve responses to their demands. The investors in Lamech and Saeed's study came from North America, Western Europe, and to a lesser extent East Asia, Japan, and Africa. Most of them reported being satisfied with their investments in small systems in the Dominican Republic, Kenya, and Nicaragua. Many more reported being satisfied and very satisfied than dissatisfied with their investments in El Salvador and Peru. Most reported being satisfied to very satisfied, that is, they considered future investments in Bolivia, Costa Rica, Guatemala, Jamaica, Morocco, and Panama.

Every investor knows Warren Buffett's number one rule: "Never lose money", and they also know his second rule: "Never forget rule number one". So, if countries want to attract money, they must tune in to investors. Likewise, startups must put themselves in the shoes of investors. Few startups receive funding, and their life expectancy is low, but despite that, they have become fashionable, how so? We will examine later why some people are betting on startups. We will also review startup success factors according to investors, founders, and researchers. But first, let's see how companies can introduce artificial intelligence to their business.

Artificial Intelligence for Companies

Some companies may ask: Why is it necessary to bring artificial intelligence experts on board if we are millennial sailors? Such an entity may know its sea better than anyone else. Still, it is wise to recognize the transformative power of intelligent technologies to propel a vessel before an artificial tidal wave sinks it. The question is not *whether* an organization will be affected, but rather *when*. Depending on each organization's size, it is possible to integrate experts into your team or acquire the services of a specialized consulting firm and strategically train your staff. For companies, finding the right experts can be as tricky as finding a "needle in a haystack". If a company is not familiar with the scope of cutting-edge technologies and what should be the profiles of professionals to lead artificial intelligence projects, they can turn to specialized recruitment agencies, their network of contacts, universities, or research institutes. Alternatively, organizations may choose to hire specialized consulting firms. For example, consulting firms Landing AI, PricewaterhouseCoopers, and Xomnia approach the problem of how to introduce artificial intelligence in an organization, each with different recommendations as shown in Table 2.4.

Renowned AI researcher and entrepreneur Andrew Ng called artificial intelligence the "new electricity" of this technological revolution. Ng and computer scientist Daphne Koller, founded Coursera, an online education platform. In 2017, Ng founded the consulting firm Landing AI. Landing AI's proposal to "electrify yourself" with artificial intelligence has five steps as well as the proposal of the multinational consulting firm PricewaterhouseCoopers. The suggestion of Dutch consulting firm Xomnia offers four alternatives. There will be many other consulting firms offering the same service in the market, but we will only review these three proposals.

Looking at Landing AI's proposal, it is clear that from the very first step, consultants recommend involving AI experts from outside the business. The path to intelligent electrification takes more than half a year. First, the consultant firm is responsible for developing a prototype, which can serve as a proof of concept to convince business owners of artificial intelligence's benefits. The next steps are to put together an internal team, provide training—possibly through Deeplearning.ai's online courses, some taught by Andrew Ng himself, plus other courses from Coursera—and then, develop a strategy over your internal and external communications.

The proposal by Christian Kirschniak, a business expert at PricewaterhouseCoopers consulting firm, also suggests following five steps

Landing AI suggests:
1. Setup demonstration projects together with external artificial intelligence experts to convince yourself that artificial intelligence works (6 to 12 months).
2. Put together an internal artificial intelligence team.
3. Facilitate training in artificial intelligence (4 hours per executive, 12 hours per project leader, and 100 hours per engineer).
4. After acquiring training and experience with artificial intelligence, develop your strategy.
5. Expand your communications: internal (talent) and external (investors, government, customers), to explain how artificial intelligence will affect your business.
PricewaterhouseCoopers suggests:
1. Know how artificial intelligence will affect your business.
2. Recognize your strengths, find out if you have available datasets.
3. Create an artificial intelligence strategy, knowing how artificial intelligence will reinforce your business and what business models, products or services are possible.
4. Make sure you are ethical and transparent.
5. Hire artificial intelligence specialists and train your employees.
Xomnia suggests:
Choose from one of the following four options:
a) Let consultants develop a complete project in two phases: the first based on the definition of a practical case, the second based on the project's execution.
b) Put yourself in the hands of consultants.
c) Let an artificial intelligence junior work four days in your company and one day in Xomnia to get training and guidance for one year.
d) Take the training courses offered by Xomnia consultants.

Table 2.4: How to adopt artificial intelligence in an entity, according to consulting firms Landing AI, PricewaterhouseCoopers, and Xomnia. Based on LandingAI (2020); Kirschniak (2018); Xomnia (2014).

to becoming "AI champions", without offering a time horizon. He proposes as a first step to familiarize yourself with artificial intelligence, recognize your business strengths and available datasets, develop a strategy ensuring ethical and transparent behavior. Finally, in the fifth step, he suggests hiring AI specialists. Possibly all actions under the guidance of consultants.

Xomnia consulting firm, founded in 2013, offers four solutions. In the first alternative, Xomnia's team defines together with the client, a practical case to be then executed. In a second alternative, they suggest a consultancy; in the third alternative, they offer the possibility that a junior professional is trained for a year in your company with guidance and support from consultants. The last alternative, like the other two consulting firms, offers training courses for your company.

Which of the three consulting firms' proposals works best to get smartly electrified? Certainly, there are no empirical data to compare them. The solution with the best results will surely be the one that has AI experts from the start, has the necessary talent to implement the solution, quality data exists, appropriate infrastructure is in place, and local regulations are friendly. Companies will have to be careful when choosing their people because, given the demand and shortage of qualified AI professionals, many false prophets will appear, and companies will risk acquiring artificial stupidity technologies.

Imagine that a client asks a translation agency for a text in a language that she or he does not speak. If the translation agency is not certified by a widely accepted authority, the client will have no way of knowing whether the translation received is good or not, unless a credentialed auditor confirms the translation's quality. Recovering money given to a phony is almost impossible. The issue of training and auditing in artificial intelligence is vital. Likewise, there will be dissatisfied customers who cannot appreciate good work. Professionals will have to contemplate that possibility as well.

What minimum knowledge should an artificial intelligence professional have? It all depends on what kind of work he or she is going to perform. Similarly, we might request a minimum knowledge level for a person to build a chair, a house, or a suspension bridge that guarantees safety. The essential requirement for artificial intelligence inventions lies in the people in charge and the control mechanisms that ensure proper development. A five-year university study in an artificial intelligence program or its equivalent should guarantee that a person knows about this field, but in many cases, specialization in this area is acquired only with Ph.D. study.

However, as demand for AI-savvy professionals grows, those with degrees in the exact sciences and engineering are opting to take specialized AI courses to catch the train and become certified. The challenge in the coming years will be for customers to spot discredited developers who offer toy prototypes that appear to work, but in the real world fail miserably or, worse, can cause damage. If a house collapses, the professionals who built it are responsible for the damage. How much damage or harm can cause a biased suggestion to qualify an applicant for a job? What impact has credit denial on a person? What repercussions do the erroneous prediction of a system have in courts of justice when sentences are handed down? One of the principles of The Organisation for Economic Co-operation and Development states that: "Organisations and individuals developing, deploying or operating AI systems should be held accountable for their proper functioning". Accidents do happen, but there are rules for dealing with them. Ideally, maximum security should be ensured, in addition to covering issues such as privacy, bias, and ethics.

Minimum Requirements for AI Projects

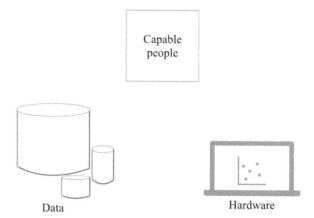

Figure 2.3: Minimum requirements for artificial intelligence projects: capable people, data and hardware.

The minimum requirements to execute artificial intelligence projects are:

- A capable team,

- Datasets, and

- Computers with software and documentation.

Of the three components of Figure 2.3, the most important element is to have a capable team. However, in artificial intelligence, the following phenomenon occurs: at some point, the amount of data and its quality begin to take center stage along with the hardware to process it. A good team of professionals with excellent data can provide better results than a team of experts with little and poor-quality data. Hardware is relevant because some algorithms do not run if they are not processed on particular computers.

We can compare artificial intelligence projects to a Formula One competition. To win the race, you need a competent team with skilled racers, but the vehicle you are going to race in, is decisive for the team. A Fangio or a Schumacher could not win a competition sitting in an old nonracing car, while a good racer could beat them both if he or she is at the wheel of a state-of-the-art racing car. The best thing that could happen to a nonprofessional Formula One racer would be to lose the race because they would either be killed in the attempt or cause a serious accident. Likewise, artificial intelligence projects require competent people to execute projects, but they depend on the quality and quantity of data at their disposal and the appropriate hardware.

A good Internet connection is not essential but relevant during the creative process. It is also necessary to have up-to-date documentation at all times and to be always up to date with advances in the field, with access to state-of-the-art publications. The greater the computing power, quantity and quality of data, and expertise of the group of people working on a prototype or project, the better the results will be.

A prototype is understood as a model that demonstrates how a product or service works. It can be used to prove a concept. For example, can we use an algorithm to recognize digits automatically? Yes, indeed the neural network, considered by Makridakis as the trigger of the artificial intelligence revolution, was designed to recognize handwritten digits; very useful in the automatic recognition of postal codes on letters, numbers on checks, amounts on account balances and any number that is visible on a surface. The legendary dataset used in this type of application is the MNIST by LeCun, Cortes, and Burges, which contains 60 000 examples to train models and 10 000 examples to test them.

State-of-the-art scientific articles describe the datasets used for their research and often even reference the repository from which the data can be obtained. Also, datasets can be obtained from different sites, often freely. For example, Deeplearning.net has a list of datasets cataloged in areas such as music, natural images, artificial sets, faces, text, language, recommender systems, and miscellaneous; also the Wikipedia encyclopedia has links to datasets of images, text, sound, signals, physical data, biological data, anomalous data, questions and answers, multivariate data, and what they have called curated repositories. A beginners' "fantastic idea" is to be willing to create their own dataset. While creating datasets is a task that helps many researchers and developers, this should be considered *only* if there are no other datasets available to test a hypothesis. Moreover, expert researchers generally discard projects if data to execute a project does not exist.

Minimum hardware features for machine learning algorithms, other than deep learning, are a topic that seems to be of little concern. For example, Scikit-learn can be used to develop machine learning algorithms and is an excellent open-source tool.

Powerful computers are needed to develop deep learning algorithms. Google Colab helps you avoid the specialized task of installing and properly configuring expensive GPU servers for deep learning. Google Colab offers these GPU servers preinstalled and ready to bake deep learning algorithms for free while used for education and research. One of the most popular programming languages for developing artificial intelligence prototypes is Python, and there are several tools around it, Numpy, Matplotlib, Pandas, Scikit-learn, TensorFlow, Keras, OpenCV, OpenAI Gym, and many more. Others prefer to work with R, C++, or Matlab.

In addition to the classic books in this field, mentioned in the first chapter, publications are another vital source of knowledge. A trend in research is to publish state-of-the-art results on open platforms such as Arxiv even before going through the long path of peer review, but in that case, critical expertise is necessary to recognize the quality of each publication. It is also possible to freely access lectures by eminent scholars on Internet platforms such as Videolectures.net, Videos.neurips.cc or YouTube.com.

The Fascination for Startups

What is the magic behind technological ventures? How many startups in general die, survive, or gloriously succeed?

Startups are technological ventures that promise slow and sudden, accelerated growth in a relatively short period of time. Although the life expectancy of startups is low, they have recently been given quite a lot of coverage. Startups are like a lottery, not only for the founders but also for the countries that host them. The reward is very high, but it is not guaranteed.

The trend of institutionally supporting startups is relatively recent. Many countries have taken on the task of creating a proper environment to promote startups. They support the establishment of entrepreneurship ecosystems through promoting policies, startup clubs, classes, incubators, entrepreneurship and innovation programs in universities and outside of them. Participants are bitten by the entrepreneurial mosquito, receive advice on business models, intellectual property, training to present a prototype or business model, legal and accounting advice, as well as contact with mentors and investors. Many universities have technology transfer offices also interested in commercially exploiting inventions coming out of academia.

Ko and An explained in their 2019 article that to combat youth unemployment in Korea, which had reached 10 percent in 2008, the Korean government promoted the creation of startups, and their visibility in the media increased by about eight times over a decade. By 2017 there were more than 800 programs supporting startups. Anyway, Ko and An observed that few startups survive their first years of life, and the measure of success of a startup is given by its ability to stay in the market by its "sales after three years of operations". Although the support of the government or university does not guarantee the success of startups, Ko and An found that the success factors of startups have to do with:

- the founders, their "independence, vision, and leadership",

- how flexible the startups are with regard to their business model toward the markets where they can generate sales,

- the team, its know-how and experience, resources that are scarce in young startups, but can be compensated with a team whose generations converge, and

- excellent mentoring, associated with the mentors' ability to address problems. Although mentors may affect the original startup's idea, mentors can help direct operations, business model, and resources that the startup can acquire.

In 2019, Prohorovs et al. published a study conducted in Latvia and Russia, in which they observed that startups with large success potential are the ones that attract professional investors. The rest are financed through different schemes, for example, scholarships, crowd-founding platforms, or the well-known financing type FFF: "family, friends and fools". They also explain that although attracting capital is essential for the survival and growth of startups, only two percent of startups in the USA receive funding from venture capitalists, and once a startup gets funding, it is very likely that an external manager will replace the founders.

The study by Prohorovs and his colleagues focused on discovering what makes a startup attract capital, and in a ranking of six priorities, they found that indeed, investors and entrepreneurs agree that the most important thing for the success of a startup and its potential to attract investment is the "Management skills of the founders and the team", but then they diverge in their opinions. Investors are quite interested in the market's size and how it is growing, the competition, the startup's financial potential and how scalable the product is; to a lesser extent they worry about whether the founders are reliable or whether they have previous experience in entrepreneurship, while entrepreneurs consider that after their own management skills, the most important thing to succeed and obtain capital, apart from the characteristics of the product, is their training, how specialized they are, how much previous experience

"Startup success factors according to investors"	"Startup success factors according to entrepreneurs"
"Management skills of the founders and the team"	"Management skills of the founders and the team"
"Market size and growth rate, competition"	"Product characteristics and scalability"; "Specialized education and skills"
"Financial potential (business plan, profitability of the project, exit opportunities)"	"Entrepreneurs' capital availability"; "Founder's previous experience"
"Product characteristics and scalability"	"Market size and growth rate, competition"; "Direct communication with business angels and investors"
"Founder's trustworthiness and reliability"	"Founder's trustworthiness and reliability"; "Managerial support"
"Founder's previous experience"	"Marketing and sales skills"; "Financial potential (objectives, investment target)"

Table 2.5: Startup success factors that influence capital acquisition, according to investors (literature review) and entrepreneurs (literature review and interviews), in descendent order of importance from highest to lowest, as presented in the analysis by Prohorovs et al. (2019).

they bring to the table, how much capital they have available and if they have a direct communication with investors or business angels. Then they look after the market and their competition, but curiously and contrary to the investors' ranking, entrepreneurs give the least importance to the startup's financial potential, see Table 2.5.

The study results in Latvia and Russia tell us that founders live for the illusion of their business and investors for the illusion of making money. In any case, how entrepreneurs face innovation seems to be related to the entrepreneurs' nature and their environment.

The legendary ecosystem of Silicon Valley in the USA is a world reference point in entrepreneurship and technological innovation, known for producing the most exuberant companies of recent times. Silicon Valley's ecosystem is known for being vibrant, culturally diverse, well-connected to renowned universities, established technology companies, and startups. It also possesses an excellent network of investor connections and entrepreneur-friendly policies.

But, new suns are emerging on the horizon. Already since 2006, economist Sandeep Kapur observed the progress in the software industry in countries that have a relatively cheap labor force skilled in information technology. He stressed that the advantage for India, Ireland, and Israel is the English language that allows them to easily export software development and technical support internationally, while until then, Brazil and China had grown in response to the needs of local companies.

Currently, China has set itself to lead in artificial intelligence by 2030 and has taken it very seriously. Jing and Lee explain that technology centers in China are spread across Beijing, Shenzhen, and Shanghai. Their government allocated more than 20 percent of its national social security funds for venture and private equity investments, creating more than 1 600 incubators for high-tech startups. They also created several programs to study artificial intelligence. Giants like Tencent, Alibaba, Baidu, or Xiaomi were born in China. China has technology fever, and companies are growing at a dizzying rate. K.-F. Lee believes that while Silicon Valley entrepreneurs work fiercely following a sublime mission and penalize copying, entrepreneurs in China, for whom copying is part of their culture, are capable of doing anything to satisfy the market's appetite. Lee notes that in war and startups, everything is valid.

India is another important center, which has some 3.9 million people working in information technology. A 2020 VICE News report presented by Krishna Andavolu shows the positive expectation about the future of India. It is believed that the center of the world is moving east; India has more than 5 000 registered startups, computer science studies focus also on entrepreneurship so that technologists know how to sell their ideas or prototypes and develop a business vision. As shown in the report, the interviewees returned from the USA to India because of the size of their country, as the massive Indian population makes it possible for an innovation to grow quickly, "the rules are being made", India is cosmopolitan, and entrepreneurs feel very comfortable when they return. Furthermore, the report shows that local students no longer have the goal of emigrating to the USA, but instead, there is a feeling of contributing to their country.

Russia has its Skolkovo Innovation Center built a few years ago in Moscow district, and according to its 2019 report, it is open for international cooperation. It has more than 1 900 startups operating, which generate more than 30 000 jobs and have produced more than 2 000 patents. Moreover, the Skolkovo technology ecosystem has connections to investors, a hundred industrial enterprises, 45 research centers, and other educational establishments with 85 professors and about 600 students.

According to researcher Marina Pasquali, nearly 12 percent of working-age people were engaged in startups in the Latin American and Caribbean region in 2017, and 16 percent of the ventures were founded in Brazil, nearly 12 percent in Guatemala, with the finance sector being the most common among ventures. Also, almost 2 000 startups went through the StartupChile program, and nearly half of them are active. In 2019 Redacción Economía y EFE reported that Colombia created the Center for the Fourth Industrial Revolution, which focuses on artificial intelligence, the Internet of Things, and blockchains.

The Disrupt Africa report considers 2019 to be a record year for investment in Africa: 311 startups received some USD 492 million in investments, mainly in Kenya and Nigeria. Egypt is the country that secured more investment than any other North African country, while in Southern Africa, although with less accelerated growth, technological entrepreneurship was also observed in 15 countries, including Uganda and Ghana, with the finance sector being the most popular sector, as in Latin America and the Caribbean, but also with considerable activity in the following sectors: transport, logistics, health, e-commerce, and agriculture.

Startups are a promise with no sure reward, almost like a lottery game. Startup founders generally work longer hours than any employee, and they also work days off; and when a startup is funded, investors at some point replace the startup founders with people with more experience in business management. A stable job, on the other hand, allows employees to concentrate on a specific job for which they specialized without having to worry about the challenges that startups face with financing issues, unfavorable policies, legal and fiscal issues, and the constant search to find the business model that will lead them to success. Anyway and in some cases, the temptation to run the risk of founding a startup is appealing.

Some successful startups were born out of long-term workplace friendships, where coworkers managed to get along, knew each other's qualities, and learned about a business's nuances before deciding to startup. This seems to be the case of WhatsApp's founders: Brian Acton and Jan Koum, who worked for Yahoo for several years, and after resigning, spending some time traveling, and being rejected by Facebook when they applied to work as employees, dedicated themselves to developing WhatsApp, which according to Olson, was acquired by Facebook for USD 19 billion; a sum that perhaps Facebook would have saved if it had hired both of them when they applied as employees.

Magic Triangles and Their Pacts

While the entrepreneurial landscape sounds highly intimidating, people's creative abilities are limitless, genius can be born in any corner of the planet, and the fluidity of digitalization powered by artificial intelligence can make the impossible possible.

What should be the ideal setup of student startups to harness various capacities and enhance them with artificial intelligence? Startup incubators usually form ternary groups in which people with different skills are matched to complement each other and to diversify the applications to be developed.

We can expect that people specialized in artificial intelligence do not know exactly what challenges doctors, psychologists, economists, or civil engineers face. At the same time, professionals from diverse areas do not master artificial intelligence either. By matching students from different areas with artificial intelligence students, it might be possible to create exciting and diverse solutions. At the end of the day, it will be necessary to know about business to commercialize a developed innovation.

Many universities require students to complete a number of practical hours in a company; alternatively, universities are encouraging students to participate in an entrepreneurship and innovation incubator program, promoting interdisciplinary groups to develop a business plan plus a prototype or minimum viable product. Incubators generally offer accounting services, legal advice, intellectual property advice, connection with mentors and investors.

Figure 2.4 schematically presents a ternary team proposal with capabilities in artificial intelligence, business, and "wild card". A "wild card" can be exchanged by a person in any study area, such as medicine, psychology, law, politics, environment, marketing, journalism, petroleum engineering, architecture, physics, chemistry, art, sports, or any other branch. The interdisciplinary proposal is supported by the fact that interdisciplinary teams are known to outperform homogeneous teams over time, especially after the "brainstorming" phase, as evidenced by the 4.0 Toolkit for Leaders developed by the World Economic Forum, which found that diverse teams are more profitable, have better returns on investment, are more innovative, are 30 percent better at making decisions and detecting business risks. However, other startup configurations may achieve better performance than the one shown in Figure 2.4. A specific startup study would be necessary to confirm this.

Members of a startup must know about business plans, financial projections, and intellectual property nuances. Additionally, they must understand that internal conflicts can occur, and without a partnership agreement, a friendship can turn into a duel to the death.

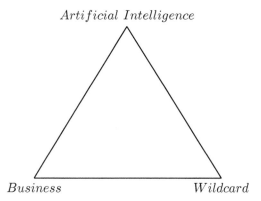

Figure 2.4: Basic formation of interdisciplinary teams in artificial intelligence incubators. (Velarde, 2020a). Reprint courtesy of Gissel Velarde.

A partnership agreement is a written document that a startup's participants make before they start working on a project. According to Helena Hernández, the partnership agreement usually have: first, general clauses regarding the agreement's duration and objectives; second, control clauses to define decision-making and responsibilities; third, protection clauses, stating participation and percentages, non-competition, confidentiality and exclusivity issues; and finally, exit clauses, explaining actions to be taken in case the company ceases to exist or a participant exits the agreement.

Without a partnership agreement, unpleasant scenarios can occur. Lawyer Josep Navajo explains some complicated situations that can happen to startup founders. For example, imagine two partners verbally agree to work on a project in equal parts. One agrees to contribute money, and the other commits to programming. After three months, the programmer did not work on the agreed code and decides to abandon the project. As the company becomes diluted, should the partner who invested capital get half of the investment? If they did not regulate their responsibilities during the project, the question of whether the remaining capital should be split remains open. Another example is the following. A startup generates sales with a product. One of the partners simultaneously opens another company with a similar product. The judgment on whether the parallel business is considered competition or not is subjective, as long as no limits were defined in a partnership agreement. Each partner can interpret the opening of the parallel business at their convenience. In real life, even with a well-defined partnership agreement, personal interpretations can take on different nuances.

High-Performance Teams

Machu Picchu, the pyramids of Egypt, or the Great Wall of China were built with the vision, skill, and sweat of the people. For all these projects, there were visionaries who had the construction's idea, experts who solved the technical problems, so that the constructions transcend time, and people who built these monuments with their own hands. Today, robots may take the heaviest part and free us to concentrate on more creative tasks. But for this to happen, it is necessary to prepare the path.

The color palette is infinite, as is the variety of applications that can be "electrified" with artificial intelligence. Usually, innovations that eat up the world are those that solve a problem, an intense pain, or take a stone out of the shoe. By solving a problem of their own, some entrepreneurs solve the affliction of many. For example, Marcelo Guital, who sold his Chilean company Agua Benedictinos to Coca-Cola for more than USD 5 million, is an entrepreneur who confessed in an interview with Revista Capital, to seek solutions to these intense pains. But some visionaries create a need or dependency for millions, which before their appearance did not exist, as was Steve Jobs's style.

Excellent project management can ensure that tasks are successfully completed according to certain evaluation criteria. Agile methodologies, budget and portfolio management, milestone tracking, and collaborative tools are all means that can help achieve project objectives.

What do high-performance teams do? How do they create collective knowledge? In an interview, Mario Moussa, management consultant, talks about the book he wrote with his colleagues Madeline Boyer and Derek Newberry: *Committed Teams*, and explains that high-performance teams have excellent communication and interaction among team members and constantly repeat three activities: first, they define their "goals, roles and norms", second, they review their "goals, roles and norms" from time to time, and third, they make adjustments between their performance and their "goals, roles and norms". Furthermore, I might add, high-performing teams know what they are doing even if they do not know all the answers to the problems they face and have a voracious appetite to achieve their goal.

Artificial intelligence can be applied in almost any field. A careful project selection is fundamental. Consider if a project is well defined and contributes to achieving the objectives of your strategy, if there are experts to lead the project, if the necessary resources can be acquired, and if the project promises a good return on investment with scalable growth possibilities. Once a project is chosen, the recipe says: "analyze, execute, experiment, and evaluate".

Chapter 3

The Laws of the Technological Jungle

Sometimes we think that the laws of the jungle are just harsh. "Kill so you don't get killed" or "the big one eats the little one". But the jungle is full of secrets and surprises. Not necessarily the biggest one is the one that survives, and appearances can be deceiving. Without the cooperation that exists between organisms, the jungle would not be the wonderful place it is.

In the jungle of life, you can find a little bit of everything over the course of your expedition. We enter into the depths of the jungle without knowing what awaits us. Ideally, one should count on the experience of someone who already knows the charms and dangers of the wild. In any case, the jungle is so large and deep that not even the most experienced traveler will know about all waterfalls, swamps, fruits, and snakes.

As an engineer and computer scientist, I have met many wonderful people during my twenty-year experience in this field, and I have been privileged to work in industrial and academic environments. I think the system has many paradoxes. Although my training allows me to understand algorithms better than I can understand people,

I have cultivated an intense curiosity to better understand people and their environments, especially in technology. I hope sociologists, psychologists, anthropologists, economists, and biologists will be tolerant of me, in addition to technologists, of course.

In the communities where I have been active, I have typically been part of one or more of the minorities. For some years now, the thriving scientific community I belong to has been trying to understand why some women who join the community, suddenly disappear, and how to promote greater diversity. I think diversity is interesting not only for artificial intelligence and technology communities or businesses but in general for men, women, and those who do not identify themselves with these categories as well.

There are several reasons why technology communities would like to be more diverse: to be more productive, to avoid biased solutions, and due to ethical reasons. I believe the problem of diversity is complex, and I found it to be more complicated than I expected. But, the good news is that people are working to fix the leaks. This chapter focuses on understanding why the issue of gender diversity is chameleon-like. I have been following the steps of researchers working on this topic, yet some of my assumptions and conclusions may be biased or simply wrong.

The Technological Environment

In 2020, a virtual version of the International Society for Music Information Retrieval (ISMIR) conference was held for the first time. The group in charge of promoting diversity announced that I was named "Notable Woman in the Field" and organized a session. I was pleased with the news, and invited the virtual assistants to meet me and talk about two topics. One of them was about how to address diversity if some early-career people feel discriminated against, instead of being supported, now that several campaigns have been activated to promote female participation in technology. I raised this topic because of a couple of comments I'd heard. Although, in general,

Figure 3.1: ISMIR 2018 Unconference. September 25, 2018, at Télécom Paris Tech. Photograph by Geoffroy Peeters. Courtesy of Geoffroy Peeters, Co-Chair of the organizing committee, ISMIR 2018 conference.

technologists want to encourage greater diversity. The photograph in Figure 3.1, reflects a common scenario in any of the international artificial intelligence conferences today. Geoffroy Peeters took this photograph during the ISMIR conference held in Paris in 2018 in a session called "Unconference", and I happen to be standing with my arms crossed, in a light-colored sweater, almost in the center of the picture.

The first time I attended a ISMIR conferences was in Utrecht in 2010, following the advice of Tillman Weyde, my mentor and who would officially become my second doctoral supervisor a few years later. At that time, there was no support program like those organized nowadays by the team that promotes diversity to potential members of this scientific community.

In 2006, I came to Germany from Bolivia, funded by a German Academic Exchange Service (DAAD) scholarship, for which I applied twice. Roberto Carranza Estivariz (1939–2005), my thesis supervisor in the five-year *licenciatura*'s program in Systems Engineering at the Universidad Católica Boliviana, advised me to learn German and apply for the scholarship; he believed I could become a researcher in any field. I only managed to tell him that I was rejected the first time.

In the master's program at the Südwestfallen University of Applied Sciences in Soest, I met a selected international group of students. The first year, we were about eight students who came with the same program, and from countries somewhat similar and somewhat different to mine. In all, I think there were about twenty students, of whom I was the only girl in the master's program in Electronic Systems and Engineering Management living on campus. In one of the ten lectures I took, there was another girl from Asia, always in a hurry to catch her train. Little by little, I met other girls with whom we became friends, even though they were not enrolled in the same program I was. With some of my classmates from the master's program, I established very good friendships, almost like those I created with my university classmates back home. For me, it was not new to spend university lectures in courses where a minority of us were girls. What is more, I used to get along very well with the guys we did projects with together. But I remember that the first months after my arrival in Soest were the hardest. I have never experienced a winter like that, with few hours of dim light. I have never had such a hard time learning, as I had in that master's degree. I thought that was how Samson would have felt when he had his hair cut. The bright summer was good for me.

My master's project was the hook to stay and work, on a temporary contract with Miebach GmbH, Welding Machines division, where one of my tasks was to analyze the signals from these machines, which were about the size of a two-storey apartment. I also took care of their relational database systems. In 2010, months before the end of my contract, I asked for a few holidays and went by train to the ISMIR conference in Utrecht, where I discovered a thriving scientific community in which people were talking about technology and music, and my experience was unforgettable. I got married in the spring and was a few months pregnant, and although at times I felt exhausted during the conference, I was delighted to attend the marathon conference program. In a fortunate conversation, I met David Meredith, who would later become my first Ph.D. supervisor in Computer Science and Engineering at Aalborg University in Denmark. I was doubly fortunate to have Tillman Weyde as my second doctoral advisor. Without the smooth Danish system, and above all, the excellent support of its childcare staff, I cannot imagine how I would have ended up with a Ph.D. as a member of an immigrant family with two young girls, the second of whom was born in Denmark. From time to time, I was also able to bring my mom to help me but always under the Immigration Office's restrictions along with the flight ticket prices.

In 2000, the first conference organized by ISMIR took place, and the machine learning component has become essential. Hu et al. published in 2016 a study on gender participation in this community. In the fifteen years of information on publication patterns, the researchers found that 74 percent of the scientific articles were not written in collaboration with female researchers. Besides, only 13 percent of publications had a woman as lead author.

In 2016, the ISMIR conference was held in New York, and a mentoring program of the *Women in Music Information Retrieval* (WiMIR) was established to promote gender diversity. The gender statistics from the ISMIR publications presented in Hu et al.'s study could well reflect the number of female attendees at that conference. One panel session was dedicated to discussing measures to promote female participation at ISMIR. The panelists explained that one action they had taken was to offer a scholarship or incentive dedicated to women only. Although I do not remember what the action was, I clearly remember one participant sitting a few chairs away from me, commenting to the fellow next to him that he did not understand why this incentive would be given to females only. He felt it was unfair the organizers did not offer incentives or scholarships to all early-career researchers (regardless of gender). It was hard not to have heard the comment. The voice was loud, and I was sitting just a few chairs away. The person who made that comment was not blind. I kept thinking about it.

During the conference breaks, I engaged in interesting conversations on the subject of diversity with some colleagues. Is it that the system does not give women a chance to emerge, or is it that women have no interest in eating up the world like men do? Is it a matter of expectations about the roles of women and men within society, or is it really biology and nature between the sexes that cause wage and representation gaps to exist in the world of science and business?

Where Are the Female Technologists?

In other international artificial intelligence conferences such as Neural Information Processing Systems (NeurIPS), International Conference on Machine Learning (ICML), or International Conference on Learning Representations (ICLR), 88 percent of researchers who

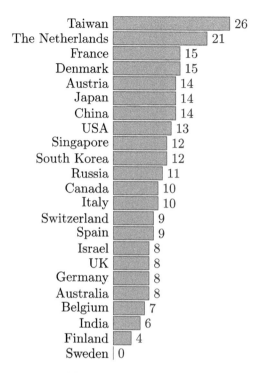

Figure 3.2: Percentage of female researchers by country who published at NeurIPS, ICML, and ICLR conferences in 2017 (out of a total of 4 000 researchers). Based on data from Mantha and Hudson (2018).

published in 2017 were male, according to Mantha and Hudson, see Figure 3.2. However, UNESCO Institute for Statistics states that in some nations, the number of female graduates in Information and Communication Technologies is higher or at least equal to the number of male graduates, as is the case in Andorra, Oman, Myanmar, the United Arab Emirates, Syria, Tunisia, Algeria, Qatar, Bahrain, Peru, or India.

Researcher Roli Varma conducted a survey-based study in 2009 to understand why girls in India study computer science, and observed that the girls' decision was influenced by the perception of their father or brothers, who encouraged them to study computer science as a choice considered "good for girls" that requires no physical effort, only intellectual and can be performed from the comfort of a desk.

A 2017 study by Raghuram and colleagues, reported that some 3.9 million people work in Information Technology and Business Process Management in India, a sector that has grown rapidly in recent years. The percentage of women working in this field has also snowballed. After agriculture, where two-thirds of workers are women, the second sector with the highest number of female workers is technology, accounting for one-third of the workforce. The study evaluated 55 technology companies and found that while just over half of entry-level recruits are women, only a quarter of management positions are headed by female managers. The highest ranks, the C-Suite, are almost entirely reserved for men, as represented in Figure 3.3.

Raghuram and her colleagues' report also shows that although females start their careers at the same age as males, they remain in lower positions, and few reach managerial positions; often, their bosses are younger than them even if they have similar qualifications. The career of women technologists in India is a short one. Most of them are single and under the age of 30. Many human resource managers interviewed are discouraged from considering female technologists because of government regulations on working hours and maternity leave. The managers interviewed believe that women with children will not be able to balance work and family responsibilities, while parenthood is not an issue for managers or other male workers.

Pooja Choux presents "A Day In The Life of An Indian Software Engineer Intern", and shows that her day starts at 6:30 in the morning. An hour later, she is already waiting for public transport in a simple neighborhood with unpaved streets to reach the modern offices of a technology company. She arrives at her destination at 9:30 a.m., after a long journey, first in a vehicle and then on an extremely

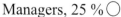

C-Suite, ○ <1%

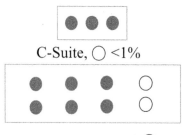

Managers, 25 % ○

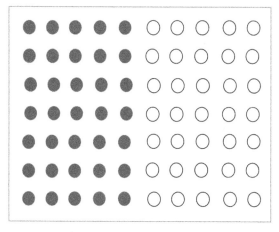

Entry level, 51% ○

Male ●   Female ○

Figure 3.3: Gender percentages vary according to hierarchy level in Indian companies in the Information Technology and Business Process Management sector. At the entry level, there is almost gender parity, a quarter of the managers are women, and the C-Suite positions are almost entirely held by men. Based on data from Raghuram et al. (2017). The graphic representation is approximate and for illustration purposes.

crowded subway. The young woman arrives at a new and elegant building where she does her internship. After passing through security checks, she takes the elevator to the eighth floor and performs her work in a spacious office with several desks. By the time she leaves her internship, it is already night. Public transport is again bursting at the seams. Arriving home at 10:15 p.m., the young woman has dinner with her parents and goes to bed to repeat the same routine the next day. If the internship schedule is similar to the work schedule, parents may not see their young children awake during workdays.

Armenia, a country that was a republic of the former Soviet Union, is another case study. Women used to be the majority in computer science universities. According to a 2006 study by Gharibyan and Gunsaulus, 75 percent of computer science students were women during the 1980s and 1990s and accounted for 44 percent of computer science students in 2006. The researchers interviewed more than 500 university students and professionals in Armenia to understand women's motivations for computer science. The interviews revealed that just over half of the Armenians (male and female) chose to study computer science because they were interested in it, and nearly 40 percent of them did so because they knew that this career would provide them with good salaries. The researchers explained that when Armenians decide to study computer science, they are persevering and resilient to difficulties and take the study with the same attitude with which they approach marriage. The divorce rate in Armenia is low compared to other countries, less than 10 percent back then.

Gharibyan and Gunsaulus knew from anecdotes that in countries of the former Soviet Union, computer science is associated with mathematics rather than engineering, and indeed, 83 percent of respondents revealed that they associated computer science with mathematics. Respondents were also asked whether computer science, mathematics, or engineering is suitable for women. A high percentage of female respondents thought that computer science and

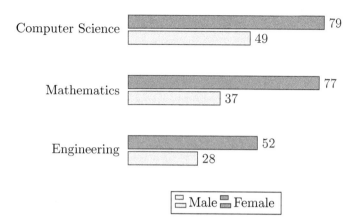

Computer Science — 79 / 49

Mathematics — 77 / 37

Engineering — 52 / 28

Male Female

Figure 3.4: Percentage of students and professionals in Armenia who believe that computer science, mathematics or engineering are suitable for women. The answers are not exclusive, as each respondent gives their opinion for each branch. Based on data from Gharibyan and Gunsaulus (2006).

mathematics were suitable for women, while only half of them believed that engineering would be. Few men considered that these three careers would be suitable for women, as shown in Figure 3.4. In addition, men said they perceived female engineers to be less attractive than female mathematicians, who are perceived to be less attractive than female computer scientists. But indeed, male respondents confessed that they would prefer to be surrounded by many pretty women in their university lectures. The researchers detected a rather curious pattern: since Armenian men are typically very attentive to women, women prefer little female competition to receive as much male attention as possible.

No open-ended questions were used for respondents to comment on why few considered engineering a suitable field for women. It could be that female and male Armenians, felt that females, despite being equally capable as males of doing intellectual work, would

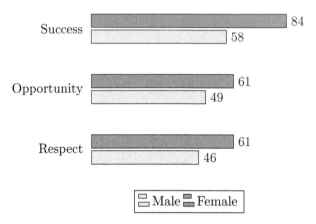

Success 84 / 58

Opportunity 61 / 49

Respect 61 / 46

Male Female

Figure 3.5: Percentage of students and professionals in Armenia who believe that computer science is a field where women receive the same recognition, opportunities and can achieve the same success as men. The answers are not mutually exclusive, as each respondent gives their opinion for each option. Based on data from Gharibyan and Gunsaulus (2006).

be at risk if they had to enter an industrial plant, and therefore consider engineering unsuitable for girls. Another reason could be that Armenians see few women in environments where engineers work and would consider such a scenario unwise for a young woman. Also, they may think that women are not capable enough to do the same intellectual work as men, or that in environments where there are few women, female supervisors are the worst tormentor for young female engineers.

The interviews revealed that although bosses perceive male professionals as "more creative and innovative", they prefer to hire female computer scientists because they are committed and loyal, choose to stay in the job they have, and are "less ambitious" than males who, once they demonstrate accomplishments, look for another job with a better salary. When asked whether Armenians believe that women receive the same respect, opportunities, and success as men, women showed a more optimistic view than men, see Figure 3.5. But it was assumed that unequal gender treatment is not specific to computing, and Armenian women were not discouraged by this. In any case, it was not asked why Armenians believe that women cannot achieve the same success as men. It could be that success is related to intellectual ability, ambition, or creativity or that women face more obstacles in the workplace than men.

Finally, another surprise of Gharibyan and Gunsaulus' study is that although it is believed that in countries such as the USA, the lack of female role models in technology affects the decision of young women not to study computer science, in Armenia, there were no such female role models either, and the interviewees did not see this lack as relevant in choosing their career. This would mean that although there were many female computer scientists in Armenia at that time, they were not recognized as female role models or did not assume positions of high visibility or hierarchy.

The Implicit Bias

Success in many cases depends on a good presentation, whether in a job interview; to close a deal with investors when presenting a business model, a prototype or a product; or to get more followers when delivering a government plan before an election. The ones who would get others to invest their time and money in them would be the charismatic speakers, who could spread emotion with their oratory, as stated in the 2019 study by Niebuhr, Tegtmeier, and Schweisfurth. The researchers developed a system to analyze acoustic parameters of voice intonation and claim that "Women sound less charismatic than men". However, there is no description of the dataset used, nor a detailed description of the software in question. At the same time, Gomez, an expert in signal processing, explains in her 2019 publication that there are several systems for voice or speech recognition that report a worse performance for female voices than for male voices.

Figure 3.6: Peahen on the left and male peacock on the right. Sketch by the author.

In the male-dominated business world, Loscocco and Bird found that women are less likely to be funded than their men competitors, as investors favor business ideas proposed by men. The question is, do female presentations sound less persuasive to investors, or are they indeed less convincing. Likewise, we might wonder if Armenian female computer scientists appear to be or actually are less creative, innovative, or ambitious than male computer scientists, as their bosses said. Is the human species like peacocks, where females do not have the same tools to impress as spectacular males do? (Figure 3.6).

If women actually sound less charismatic and persuasive, or appear to be less creative and innovative than men, it may be an effect of implicit bias. There are implicit and unconscious associations between men, rather than women, with scientific, political, or financial success or capability. Even people who consciously consider themselves unbiased about gender roles may be affected by the effect of implicit bias, see Figure 3.7 and Figure 3.8.

A 2019 study conducted by Régner, Thinus-Blanc, Netter, Schmader, and Huguet investigated whether academic excellence committees in France suffer from implicit bias since their decisions can affect thousands of top researchers' careers. More than 400 eminent academics were required to choose future leaders in various research areas, from politics, economics, chemistry, sociology, to mathematics and other fields, and in some cases, implicit gender bias was in favor of men over women to obtain scientific positions, even though committee members considered themselves as objective and rational. Régner her and colleagues noted that combating implicit bias requires educating decision-makers and teaching them that implicit bias does exist and can affect not only judges but with greater impact, those who are judged. In other words, implicit bias affects judges because it prevents them from being objective. In contrast, implicit bias affects the judged ones because their careers depend on the judges' decisions.

Harvard University's *Project Implicit* offers several implicit association tests that can be taken free of charge. Another way to convince yourself that implicit bias exists is to watch a video produced by reporters of the BBC News Mundo that shows how implicit bias is an unconscious effect that can affect people's perception, even that of those who consciously believe themselves to be impartial.

The problem of implicit bias may be reinforced not only by cultural molds but also by statistics. If 10 percent of members in technology communities are expected to be outstanding, and 12 percent are women; in a community of 100 members, only one woman may become outstanding. The same exercise can be repeated for any of the other minorities in technology or business communities.

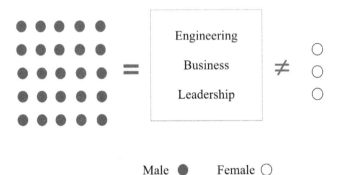

Male ●   Female ○

Figure 3.7: Implicitly positive association with engineering, business, or leadership with male gender. The graphic representation is for illustration purposes.

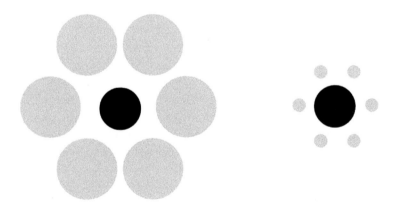

Figure 3.8: Ebbinhaus Illusion—Titchener circles. Are the black center circles the same size? In fact, they are equal, but the one on the left appears smaller than the one on the right due to the context, that is, the size of the circles around. When evaluating people, use fair measures and consider the effect of implicit bias and context. Illustration based on (Massaro & Anderson, 1971).

Corinna Cortes is one of the recognized outstanding researchers in artificial intelligence, mother of two children, and competitive runner who, together with Vladimir Vapnik, developed Support Vector Machines, one of the most popular machine learning algorithms.

The Outstanding

The first wife of one of the most legendary scientists of all times hid in the shadows for her husband to shine. Physicist Pauline Gagnon published in 2020 an investigation about the life of Mileva Marić, the invisible scientist. That story I tell below.

Letters between Albert Einstein and Mileva Marić, notes, and other documents are the evidence that both worked together developing the famous works published under the name of Albert Einstein. In addition to a thesis, five landmark publications appeared when Albert was working 8 hours a day in a patent office. If the evidence indicates that Mileva collaborated with Albert in those investigations, why did her name never appear in the publications that brought Albert to stardom?

Albert Einstein and Mileva Marić met around 1896 when they were students in the physics and mathematics section of the Zurich Polytechnic Institute and became not only a great couple but also great collaborators. Albert was characterized as "inquisitive" and "rebellious", while Mileva was a bright student with a reserved temperament. In some subjects, both obtained similar grades, but in others like applied physics or experimental work, Mileva excelled and Albert fared poorly. In a 1900 letter, Albert shows his appreciation for Mileva:

> I am also looking forward to working on our new papers. You must now continue with your investigations how proud I will be to have a little Ph.D. for a sweetheart when I remain a completely ordinary person.

Although Mileva was a dedicated student, Professor Weber failed her in her final and oral exam, and Mileva could not obtain her doctoral degree. Weber would also have been blocking Albert from getting a job at the Institute, and possibly his bad recommendation had a negative influence beyond it. Although Albert applied to several universities in Europe, none opened doors for him.

Mileva got pregnant with Albert, but his father did not want to hear about marriage as long as Albert did not get a job. Liserl, the girl born in 1902, was possibly given up for adoption. That year, Albert Einstein started working in Bern's patent office thanks to a friend's influence. The work at the patent office was very demanding. Albert worked 8 hours daily and had only one day off for the weekend. Months later, Albert and Mileva got married and Mileva took on the housework, but they both worked together in the evenings. In 1904, Mileva gave birth to their son Hans Albert.

The following year —1905— "is known as Albert's miracle year": he submitted his thesis, commented on 21 manuscripts, and five scientific articles were published under the name of Albert Einstein, although he was busy full-time at the patent office and in his private correspondence, Albert and Mileva refer to the research work as a joint effort. Two of the five articles of the "miracle year" were on Brownian motion, one on the photoelectric effect, one on special relativity, and one on the relationship between energy and mass with the famous expression $E = mc^2$. The article on the photoelectric effect won Albert the Novel Prize in 1921. The article's short-term effects describing the fundamentals of special relativity earned Albert two weeks in bed due to excessive work. In those weeks when Albert was recovering, Mileva reviewed the manuscript and submitted it. There is also evidence that some of Albert's lecture notes in Zurich were handwritten by Mileva and that patents developed by both of them did not bear her name.

In 1910, their son Eduard was born. Two years later, Albert and his cousin Elsa Löwenthal entered into a secret relationship for two years, time in which Albert worked as an academic in Prague, Zurich and finally in Berlin, close to Elsa, his new love. A couple of years later, Albert set Mileva a number of conditions that would annul her. Albert expected Mileva to be like a maid. He even forbade her to work at a desk. They divorced in 1919.

When in 1925 Mileva claimed to Albert that she deserved the Nobel Prize money and she would disclose her research's contribution, Albert responded:

> But you really gave me a good laugh when you started threatening me with your memories. Doesn't it ever dawn upon you for even a single second that no one would pay the least attention at all to your rubbish if the man with whom you are dealing had not perchance accomplished something important? When a person is a nonentity, there's nothing more to be said, but one should then be modest and shut up. I advise you to do so.

Today, Albert's attitude towards Mileva may shock us, but possibly by the standards of the time, it would have been considered reasonable for those who had the right to decide. For example, the right to vote for women was recognized in 1918 in Germany and in 1971 in Switzerland at the federal level.

I speculate —like Gagnon does— that Mileva, who had long believed that together with Albert they were a single entity: *"ein Stein"*, would have agreed with him to publish under his name because, at that time, a woman's name would have been an obstacle to publishing scientific papers. Besides, if they had been published, the scientific community may have attributed less value to them just because they were written in collaboration with a woman.

In today's world of research, one parameter used to measure the importance of a scientific article and its impact is the number of citations it has. The number of publications' citations is used to select candidates, allocate budget for research projects, or to promote researchers to higher positions. Today, according to research by Régner et al., publications by female researchers receive 10 percent fewer citations than do male publications, even if they were similar. Since 1901 and until 2018, 20 of the 607 Nobel Prize laureates in science and medicine are women, that is, about 3 percent of the prizewinners.

Albert Einstein is considered a myth, a character with a superhuman work capacity that allowed him to produce transcendental research recognized by prestigious international awards. After all, Albert and Mileva's strategy was not so bad, at least for one of them, at that time.

The Legacy of Centuries

Alicia Itatí, a doctor in education, explains that during the Roman Empire, since 27 before the Common Era, women were allowed to attend public shows, as well as to participate in politics. They had rudimentary instruction in "reading, writing and arithmetic", as well as knowledge in "oratory, history and philosophy". But after the fall of this empire, in 476 of the Common Era, the Aristotelian view that women are by nature inferior to men, and therefore must be governed by them, was maintained for centuries. Women were then destined to reproduction and forbidden to enroll in higher studies or to participate in politics.

Arthur Schopenhauer is considered by philosopher Julian Young as one of the most influential German philosophers of the nineteenth century. *The Art of Dealing with Women* is a collection of texts by Schopenhauer originally published in 1844 and 1851, that allows us to observe how the thinking of the time was regarding the role of women in Western society. For Schopenhauer, women lacked talent and sensitivity for "music, poetry or the plastic arts", even if they were good at simulating it, and he thought that:

It is men, and not women, who obtain wealth; therefore, they are not entitled to its unconditional possession, nor are they qualified to administer it.

Schopenhauer also believed it convenient to follow the tradition of the Greeks, who according to him, had a sign in the theaters saying "let the woman keep silent in the theater" and thus also put up a showy sign saying "let the woman keep silent in the assembly", a thought that does not reflect a joke by the author.

Livni and Koft explain that according to population censuses in the USA, families have typically two children per family since 1920, while in 1850, it was common for families to have 6 to 9 children. Up to that time, it was common sense to think of women as belonging to the household and men as facing the world. Livni and Koft further cite an article published around the 1850s in the San Mateo Gazette that said that a woman could "be happy in the love of her husband, her home, and its beautiful duties without asking the world for its smiles and favors".

Economist Richard Easterlin explains that until the end of the nineteenth century, in many places, women's role was focused on family reproduction, mortality was very high, such that only half of the children reached adulthood. In England, for example, women married as teenagers and entered a state of "pregnancy and child-caring" for about 15 years. This period was reduced to nearly four years by 2000, becoming a revolution for the "freedom for women", which also has to do with lower infant mortality rates and better family planning methods, with progressive effects in several countries since the end of the nineteenth century. According to Velvi Greene, an expert in public health and bacteriology, "the war against infectious disease", better housing, hygiene conditions, and better nutrition thanks to advances in agriculture, industry and technology, can be considered the pillars of "the health revolution" from the end of the nineteenth century. Also, Greene explains that the present-day hospital is a twentieth-century phenomenon.

According to Itatí's research, the University of Bologna's decree of 1377 expressly prohibited women's admission to the university. A conference held in New York in 1848, demanding women's rights, promoted the gradual and systematic entry of women into universities worldwide during the nineteenth century, especially in medical schools. This coincides with the improvement in infant mortality rates since physicians recognized that until then, many of the causes for infant mortality were basically related to the lack of knowledge of the female sex and its basic norms of hygiene.

Reduced infant mortality and effective family planning methods allowed women to enter the world of work outside home progressively in several countries from the late nineteenth century through the twentieth century, and up to the present day. In practical terms, this implies that women's entry into university helped to drive the industrial and health revolution, allowing them to participate in work outside the home. This represents a valuable contribution to the current state of development of the Artificial Era society, certainly with a more beneficial impact than thousands of years of philosophy.

Itatí also explains that during the Middle Ages, approximately between the fifth and fifteenth centuries, those in charge of engineering and construction works were male. She further describes that men exclusively performed all types of jobs in machinery and warfare weapons, and certainly, to this day, great value is attributed to engineering work; and in many places, it continues to be strongly associated with the male gender.

Bolivia, for example, is a country with equal participation between genders in the labor market. It also has one of the highest female ratios of participation in research in the world, but only about a quarter of those enrolled in engineering programs in the three most populated cities of this country are women. The 2018 study by psychologists Montenegro Castedo and Schulmeyer analyzed the differences in the labor development of more than a hundred Bolivian engineers in the fields of industrial engineering, computer engineering, systems engineering, geological engineering, petroleum engineering, and financial engineering. The researchers evaluated the resumes of 123 engineers, their specialization studies, years of work experience, absenteeism, and the number of times professionals changed positions. All male engineers held senior positions, while only 20 percent of the female participants held senior positions, and many more women than men held lower-level positions. Female engineers reported greater job stability and lower absenteeism than male engineers. But based on the comparative analysis of the profiles of Bolivian engineers, the research results showed that even when there is no difference in academic preparation between men and women, and women being more stable in their jobs than men, male engineers achieve a higher hierarchy than female engineers, similar to what happens in India. These results may indicate that subjective perception is crucial.

Popularly, female beauty is often associated with stupidity or weakness and ugliness with intelligence and ability. A Colombian soap opera that was a worldwide success was *Yo soy Betty, la fea* (*I am Betty, the Ugly*), created by Fernando Gaitán and directed by Mario Ribero Ferreira during 1999 and 2001. The soap opera portrayed Betty's life, an ungraceful and old-fashioned economist, but sweet, innocent, and very capable. Betty wore thick glasses, braces and was characterized by a funny laugh. Betty came from a simple family, dressed in the style of her grandmother's magazines, and worked in a fashion company. In the novel, the company's success depended on the work and capability of Betty, who held the position of "secretary to the CEO". In the soap opera's last episodes, Betty transforms and becomes a beautiful woman. Her boss finally falls in love with her and they get married. The soap opera's creator may have been inspired by real-life cases from the Latin American environment, but the plot caused a worldwide sensation.

Indeed, female engineers are not only at a disadvantage compared to men, but also to women, at least until a few years ago. As we have seen before, Armenian female engineers were considered less attractive than female mathematicians or computer scientists.

Something similar happened in Bolivia, when I was an engineering student —several years ago— I heard a couple of times the popular saying: "There are gorgeous, beautiful, ugly women and female engineers". Fortunately, beauty is not yet evaluated in the resumes of engineers. At least, not officially.

If statistics are still valid and Betty's plot still resonates in some environments, we could say that if you are capable, you may live in the shadows and if you are unlucky enough to be born pretty and you are an engineer, in some places you will not be taken very seriously. But if you are like ugly Betty, then you must strive to make yourself pretty, or you will not get married. Surprisingly, today, one of the largest Facebook groups in the Spanish-speaking community is the group called *Makeup tips*, which Florpeña says had more than 2 million members in 2018. After all, perhaps this is not so surprising. The second group in 2018 with more members after *Makeup tips*, also according to Florpeña, was the one called: *Healing ourselves naturally*, even with other 2 million members.

Although sayings may die out over time, engineers of the past would say with a chuckle of pride: "Architects are not macho enough to be civil engineers and not effeminate enough to be interior designers". It is also said that the road to hell is paved with good intentions. While it is true that it takes time to address some problems such as the lack of diversity or reduce gender gaps in hierarchical positions, it is also required a system allowing effective and measurable actions that are not just propaganda.

At an event called Women's Forum Mexico 2016, Mexican tycoon Carlos Slim presented emotionally his views on how "with employment and education you can solve poverty". During the talk, Slim praised the capabilities of women, recognizing that women show greater perseverance, responsibility, and even, it could be, he said, a greater ability to execute projects than men, and said that we really need to see more women in business. Slim, who is an engineer and masters numbers, showed that he knew that his company employs more than 350 thousand people, but when asked how many women occupy managerial positions in his company, he said he did not know and asked one of the two female directors of his company to remind him that there were 2 women out of 18 in key positions.

Learning from the Emerging Ones

Female representation in technical fields or management positions is greater in emerging countries than in high-income countries. After 2001 observation made by economist Jim O'Neill on the relevance of the economies associated with the BRIC bloc: Brazil, Russia, India, and China in respect to high-income countries such as Germany, Canada, France, Italy, Japan, the United Kingdom (UK), or the USA, the academic and labor progression of women in both economic blocs are often observed. For some, it is difficult to believe that emerging countries have better gender scores in some evaluations.

Figure 3.9 shows the 2017 data from UNESCO Institute for Statistics regarding the percentage of female graduates in Armenia, Germany, Brazil, Mexico, India, South Africa, UK, and the USA, in tertiary education programs in four groups: first, Information and Communication Technologies (ICT); second, Engineering, Manufacturing and Construction; third, Natural Sciences, Mathematics and Statistics; and fourth, Business, Administration, and Law. In India, there is near gender parity in the percentage of graduates in all branches except Engineering, Manufacturing and Construction. Armenia, which had trained many more female than male computer scientists three decades ago, has now reversed its balance, and also has the lowest proportion of women in Engineering and Construction. But overall, female graduates in Natural Sciences, Mathematics and Statistics, as well as in Business, Management and Law make up just over half of the graduates.

Raghuram and her colleagues report that Mexico, the Russian Federation, and Brazil are the nations with the highest female representation in the Information Technology labor sector as shown in Figure 3.10, as opposed to the UK and USA, which have the low-

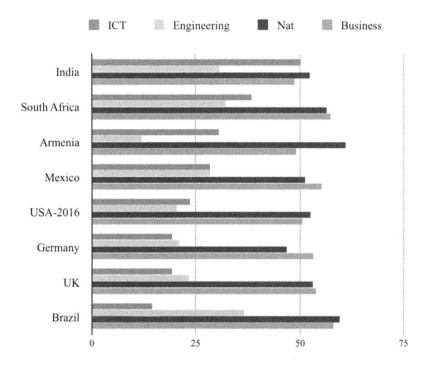

Figure 3.9: Percentage of female graduates in 2017 by country and by field of study: first, Information and Communication Technologies (ICT); second, Engineering, Manufacturing and Construction (Engineering); third, Natural Sciences, Mathematics and Statistics (Nat); and fourth, Business, Management and Law (Business). Based on data from UNESCO Institute for Statistics (n.d.). The data corresponds in all cases to 2017, except the USA data which is from 2016.

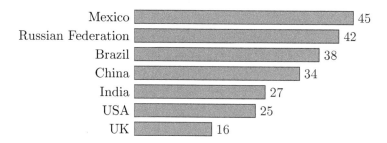

Figure 3.10: Percentage of women employed in Information Technology. Based on data by Raghuram et al. (2017).

est female representation. Finally, Mexico is the country with the highest percentage of companies in Information Technology with top female managers, as shown in Figure 3.11.

In the USA, where people dream of being their own boss, women lag behind men in small businesses growth according to a 2012 study by Loscocco and Bird, who found that businesses run by women are generally smaller and less profitable; female entrepreneurs own or obtain less capital to start their businesses, and they spend fewer hours in business because they have more family responsibilities than men. The researchers explained that culturally, women are expected to maintain the primary role in household caring, and men are expected to obtain higher-paying jobs, which influences the decisions men and women make in economic activities. Besides, Loscocco and Bird note that women, in any type of work, are more willing than men to look for ways to balance work and family, and perhaps this is one reason why women want to, or must keep their businesses smaller, local, flexible, and manageable.

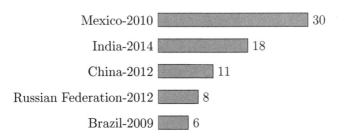

Figure 3.11: Percentage of companies in Information Technology with top female managers by country. Based on data from Raghuram et al. (2017).

But the story is not the same everywhere. A 2014 study by Cárdenas and her colleagues shows that the percentages of female participation in Latin American countries in leadership positions, including managers, senior officials, or legislators, vary from near parity, as in Panama, to only 19 percent in Peru, and in average, one-third of Latin American leaders are women.

The goal of the study by Cárdenas and her colleagues was to discover why Latin American countries have a good female representation in management positions, in general, better than that of high-income countries. Therefore, they conducted interviews with 162 highly prominent female executives in leadership positions in national and international private sector companies, which generally have high participation of female employees, in 17 Latin American countries, from Mexico to Argentina. Most female executives interviewed had university studies, mainly in business, economics, engineering, and other sciences. Most of them were married with children and had demanding jobs to which they dedicated some 54 hours a week, in addition to frequent travel. Most executives who took maternity leave returned to their same jobs after it.

Despite the perception that machismo still exists in Latin America, 38 percent of the women interviewed did not report any obstacles in their careers related to prejudice or wage discrimination, nor did they feel discriminated when it came to job promotion, nor did they report a lack of flexibility in accommodating their work schedules, nor did they face experiences of verbal or sexual harassment. The rest of the female managers did observe certain non-exclusive obstacles; 40 percent of them mentioned wage discrimination as a challenge, 27 percent observed discrimination to be promoted as an obstacle, and 19 percent mentioned other types of discrimination.

In general, female Latin American leaders were highly ambitious, satisfied, and motivated to achieve their personal goals. They revealed that the biggest challenge for them was balancing work and family. They considered their tenacity, ability, and leadership were the keys to their success rather than environmental factors. Still, they recognized as essential the support they receive from their many family members. Also, they admitted that a crucial factor in their careers depended on the low-cost domestic help available to them. Most of the interviewed leaders employed at least two types of helpers: a chauffeur or nanny and a cleaning lady, to whom they allocated less than 10 percent of their salary.

The Faces of Power

Studies show that well-managed and diverse teams far outperform homogeneous teams over time, simply because diverse teams have a broader palette of knowledge and skills. They are better at identifying problems and solutions. Besides, companies whose boards have at least 10 percent female participants, obtain up to 5 percent higher returns on equity. C-Suites that give up one-third of participation to women are 15 percent more profitable than companies with other distributions.

While it is in the best interest of companies to have more women in management positions, it is not in the best interest of women in minority environments to be supervisors. Women who are supervisors are more likely to be sexually harassed than women in lower positions, if women in a company are in the minority. This pattern has been observed in the USA, Sweden, and Japan in a 2020 study conducted by Folke, Rickne, Tanaka, and Tateishi in which thousands of workers in each country were interviewed. The results revealed that sexual harassment usually occurs when the subordinates are mostly male and even when actions can be taken against the harassers. The study by Folke and his colleagues also revealed that after a female supervisor has suffered from an experience of sexual harassment, she is subject to even greater professional and social victimization: she is labeled as a "troublemaker", she is denied the promotion or training she deserved, gossip appears, or colleagues ignore her, affecting her physical and psychological well-being and her work productivity, becoming a severe and overwhelming problem, that has been extensively documented by psychologists, sociologists, and medical doctors.

Several complaints appeared in conjunction with the #MeToo movement, and according to statistics, half of all women may face harassment at least once in their working lives through "unwelcome physical actions or offensive remarks or innuendos on subject matter that is commonly associated with sex" as Folke and his colleagues asserted in 2020, and explained that sexual harassment is not about sex, but about power through manipulation. Thus, it is a "paradox of power" that female supervisors are the most targeted.

Some organizations are adopting policies that describe codes of conduct to prevent sexual harassment or address it if it occurs. Still, Hennekam and Bennett stated that the culture embraced by companies and their employees can be a more powerful weapon in combating this problem.

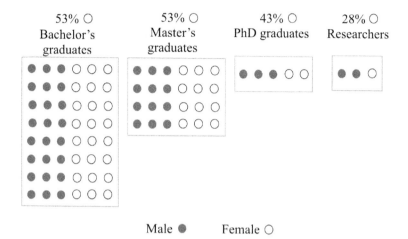

53% ○
Bachelor's
graduates

53% ○
Master's
graduates

43% ○
PhD graduates

28% ○
Researchers

Male ●   Female ○

Figure 3.12: Female participation in higher education and research, world-wide. Fifty-three percent of Bachelor's and Master's program graduates are women, as are 43 percent of doctoral program graduates, but only 28 percent of researchers in academia are women. Based on data from Huyer (2015) and United States Census Bureau (2019). To have an idea of how many students finish a Bachelor's degree, then a Master's degree, and then a Doctorate, I consulted data from the United States Census Bureau (2019), which, although it has only USA data, could offer a general picture regarding the number of graduates per program. Therefore, the graphic representation is approximate and for illustrative purposes.

Obstacles: More Complex Than It Seems

A Global Approach to the Gender Gap in Mathematical, Computing, and Natural Sciences published in 2020 and edited by Guillopé and Roy, shows that women face several obstacles such as lack of support from their relatives, they have little confidence in themselves, have no role models, may suffer repression, their careers progress slowly, they are allocated few resources and receive low pay compared to their male peers.

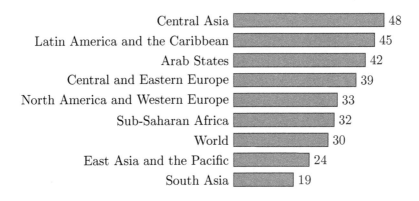

Figure 3.13: Proportion of female researchers by region. Based on data from UNESCO Institute for Statistics (2019).

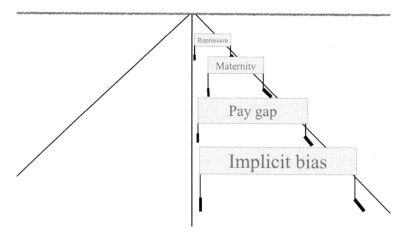

Figure 3.14: The career in the technology sector. Left lane for the majority, right lane for the minority. If the right lane increases the number of runners, obstacles may fall.

In science, the general picture is that as time goes by and people progress on training and work ladders, women begin to disappear from the scene. In a 2015 UNESCO report, Sophia Huyer recognizes this issue, which is general in higher education and science, as the problem of a "leaky pipeline". Almost 50 percent of bachelor's and master's students are women. This proportion is lower for female graduates with doctoral degrees. Worldwide, only one-third of academic researchers are women, see Figure 3.12.

However, figures vary widely by region. The 2019 report *Women in Science* by UNESCO Institute for Statistics, shows that in some regions, such as Central Asia or Latin America and the Caribbean, female and male participation in research is almost equally distributed. In other regions, female researchers are in the minority. South and West Asia present the lowest rate of female researchers (only 18 percent). Globally this figure reaches 29 percent, see Figure 3.13.

Women's career path presents several obstacles such as implicit bias and cultural molds, wage and resource gaps, maternity, and repression. Figure 3.14 illustrates the career path in technology. The outcomes of that career path may be full of surprises or paradoxes, depending on the region and environment. Because technologists have good reasons to promote diversity in their communities, incentives have been offered to future female technologists and recognition to female veterans. This can cause confusion in the majority, or even jealousy. I sense that few people know or are aware of all the disadvantages and problems women in tech face in the twenty-first century. Many women will experience the problems of discrimination or lack of recognition firsthand, but many will never be aware of the problem. It may not affect them directly, or it may even seem to them that measures promoting minorities are unfair.

The Side Effects

In August 2020, I asked in an online publication or post: "Why do few girls study engineering or computer science?" The anonymous and public comments mentioned considering history, culture, opportunity, minority intimidation, and biology. One public comment that caught my attention is the following, written by Israel Aillon (which I translated from Spanish, as):

> Reading the previous comments, it is clear that several years ago it was more complicated, but today there are too many incentives, to the point that in the software environment, there are specific and unique events to motivate women and scholarships only for them, to the point that there are school children who are confused because when there is a women's event related to technology, boys think they cannot participate. And if you search for it, you will be able to verify that aspect. A gender should not be encouraged to participate in sci-

ence issues; the new generation should be encouraged as a whole because the barriers are in our minds.

This comment makes me think that possibly the diversity promotion mechanisms currently applied are not well received. There may be several realities or different perceptions of the same reality. It also seems that more information is required, and perhaps other mechanisms are needed not only to promote diversity in technological fields in new generations but to manage to maintain diversity over time and foster diversity at all levels. Still, some may not want to leave their chair, and others may not want to take responsibility. In some ways, boys may feel discriminated against because a lot of attention has been given to girls lately. Besides, they may not understand why there is such a desire to promote women in university studies when both genders seem to manage well their studies at that stage.

Walter Terrazas also wrote a public comment to my post's question, in which he cited a 2010 series on the paradox of job segregation in high-income countries: "The Gender Equality Paradox" by comedian and sociologist Harald Eia and writer Ole-Martin Ihle. Eia interviews several experts and presents the controversy: "Culture versus Biology". In the interviews, psychologist Anne Campbell proposed that in societies where there is an opportunity for free choice, genetics will manifest. Campbell observed that in high-income countries with "greater freedom", few women are engaged in computer science or engineering, while in middle- or low-income countries like India, women would choose these fields due to the opportunity to get a better job. Another of those interviewed by Eia was clinical psychologist Simon Baron-Cohen, known for his theory that women are more empathetic and men are more systemic. In experiments with newborn babies, who have not been exposed to any culture, Baron-Cohen noticed that girls prefer to observe faces and boys prefer to observe mechanical objects. Baron-Cohen also explained that testosterone levels, which in boys are twice as high as in girls, could predict how much interest a person might have in how systems work or how much empathy they demonstrate. He considered that a balanced stance on the occupational choice between genders should consider a mix of culture and biology.

An alternative not seen in Eia and Ihle's report that comes to my mind is that people react differently to risk and opportunity. In egalitarian, high-income countries, any occupation ensures people, more or less, a comfortable life, and people will usually prefer friendly environments with low-risk probabilities. If women feel intimidated, threatened, or not so identified with male-dominated environments, they will most likely choose more female-friendly settings where they are not at risk of being discriminated against. In less egalitarian low- and middle-income countries, not all occupations ensure a comfortable life, and the differences in returns by occupation can be significant. Hence, people choose disciplines that provide a good income and promise a better future if possible. As more women enter the technological workplace, problems affecting minorities, such as gender discrimination, are reduced. Moreover, with greater diversity, the developed solutions maybe more inclusive, and in general, as studies show, productivity increases, and at the same time, more favorable environments maybe created for all.

Guillopé and Roy, who edited the 2020 report *A Global Approach to the Gender Gap in Mathematical, Computing, and Natural Sciences*, gave a series of recommendations to parents and instructors. For example, carry out activities that help girls become more confident and generally educate them about gender issues. The advice experts gave to organizations and unions has to do with fostering an atmosphere of respect, defining practices to avoid all types of discrimination, including harassment, as well as considering the impact of motherhood, and being transparent with salaries, bonuses and promotions. What I could not find are recommendations for women.

The Global Gender Gap Report 2020 by the World Economic Forum shows that men dominate technology jobs and are better paid than women, who generally work in low-paying occupations.

The report estimates that it will take 99.5 years to achieve equality between men and women in four dimensions: first, economic participation and opportunity, second, education, third, health, and fourth, political empowerment.

If a person is a woman and is aware of all that awaits her, what attitude should she take in the next 99.5 years? What should be recommended to young women in less advantaged regions? When a crisis looms, you are advised to tighten your belt, but if anticipating that conditions will be unfavorable in about a century, perhaps you should take a different attitude. However, new policies may disproportionately benefit some fortunate ones, possibly causing envy in some cases.

Some have chosen to create female technology networks where the minority are male. In some environments, this measure may help increase female participation in technological fields, but I do not know if turning over a coin can help achieve diversity.

Not Everyone Can Enter the Party

Countries with a better female representation in technology, are not in the celebration of artificial intelligence. Either they do not master the technology yet, or for some other reason they cannot get in.

The 2019 report on *Technology Trends* by World Intellectual Property Organization shows that countries with the highest number of scientific publications in artificial intelligence are China and the USA. Also, both countries have the highest number of patents, as seen in Figure 3.15, being that since 2006, China has an average annual growth of 29 percent, surpassing the USA. The patent office in China is nowadays much busier than that of the USA. The data also shows that Japan has bet on patents more than on scientific publications, although the number of scientific publications produced by Japan is similar to those of the UK, India, Germany, France, Canada, Italy, and Spain. Finally, observe that various countries do not appear with either artificial intelligence publications or patents.

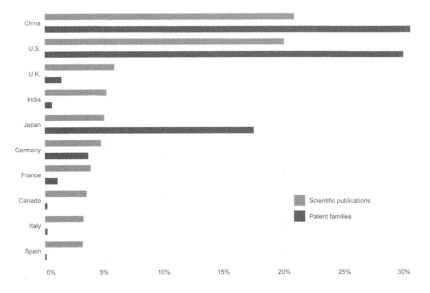

Figure 3.15: Percentage of artificial intelligence scientific publications and artificial intelligence-based patents by country. Source: World Intellectual Property Organization (2019, Fig. 5.7), CC BY 3.0 IGO.

International artificial intelligence conferences are usually held in high-income countries. Some nationalities have a super passport that opens many doors. Other nationalities are limited to enter several countries, and therefore require a visa, even if it is to attend a conference for three days. Visas are expensive, require a lot of paperwork, and to top it off, they can be refused. Some countries will offer a visa for the specific conference you are attending for the conference duration. You can also apply for a business visa granted for extended periods for you to enter and leave a country more than once under certain restrictions. I could give a whole talk about visas and anecdotes. My passports are full of visas, most of which are expired, even though I have not attended that many conferences.

For the pockets of technologists who do not come from high-income countries, attending a conference can be an expensive adventure that escapes reality. But, thanks to COVID-19, many conferences are being held virtually, and some have reduced registration fees considerably. This situation could promote greater diversity in the future. Anyway, artificial intelligence experts are geographically concentrated in certain laboratories and the truth is, if you do not know how the party works inside, it will not be easy for you to get in.

Ape Societies

Is it the environment or our genetic condition that makes us the way we are? That female participation in research varies widely by region implies that environment is a powerful factor that can predict how a person will fare at different stages of their career. The labor jungle may be very different in India, Mexico, Russia, the USA, or Denmark, despite similar patterns or trends. In some work environments, people will be more collaborative and tolerant, while in other environments, people will be more authoritarian and violent.

Studies show that minorities are at high risk of being harassed. Individuals who self-report experiencing harassment or discrimination feel very dissatisfied with their jobs, constantly think about quitting, and many do. This vicious cycle is responsible for further worsening minority segregation and deepening the wage gaps minorities suffer, as Folke and Rickne's 2020 study demonstrates.

However, even in organizations with a balanced number of males and females, there may still be large differences in how authority, tolerance, or cooperation are handled. If we look at our closest cousins' societies, we might perhaps understand what elements make bonobos harmonious and chimpanzees brutal.

According to primatologists, we share common traits with bonobos and chimpanzees. We diverged from a common ancestor millions of years ago. Bonobos developed south of Congo River, a region abundant in fruits and vegetation, while chimpanzees developed north of Congo River in a hostile territory shared with gorillas. Isabel Behncke, who has been studying apes for several years, explained in a talk that we humans resemble chimpanzees in having a dark side because these apes are brutal, power-hungry, technological, highly hierarchical, and Machiavellian. In contrast, our gentle side is similar to that of bonobos, characterized by being peaceful, very tolerant, highly collaborative, playful, and erotic. Behncke has observed that bonobo culture is led by females who, regardless of whether they are related or not, form lifelong, solid bonds with each other. Young females look for a female mentor in the group that welcomes them. In contrast, female chimpanzees are generally solitary because of the repression they suffer. Behncke further explained bonobos do not have a marked hierarchy and keep the peace, sometimes not so peacefully, but never lethally. Males in bonobo societies do not form coalitions, and the natural state of bonobo culture is play, which makes them tolerant and allows social learning based on trust, even among strangers.

Frans De Waal, also a primatologist, considers that bonobos maintain control thanks to females' solidarity. He observed that chimpanzees are patriarchal like most animals, and are brutal, competitive, and highly hierarchical such that they obey a charismatic alpha male, who stays in power thanks to the support of his friends and his fighting abilities. So the old ones cannot stay in power for long, as the female alpha chimpanzees who rule for years, are tough and keep control over young females for decades through slights and subtle aggression, using emotional games to keep them under their thumb.

Which society sounds more familiar to the reader, the bonobo or the chimpanzee? How would a chimpanzee do in the land of bonobos? Would a chimpanzee manage to adapt to the fact that its destiny depends on the patience that characterizes bonobos? Would the chimpanzee quickly understand that the laws governing bonobo society are different from those it had been used to? And how would a bonobo do in the chimpanzees' territory? Would the bonobo come out alive? The environment can shape us, but we can make decisions. You could choose what kind of society you would like to belong to. You may not be able to change your environment, but you will be able to adapt or disappear.

Improving Diversity

Diversity is a characteristic of healthy and thriving environments. As seen previously, well-managed diverse teams are known to outperform well-managed homogeneous teams over time, see Figure 3.16.

The *4.0 Toolkit for Leaders*, developed by the World Economic Forum, reports that diverse teams are more:

- profitable, between 25 and 36 percent more likely to excel on profitability,

- achieve up to 20 percent higher innovation rates,

- make better decisions,

- can better spot business risks, and

- employees are more engaged.

Similarly, Raghuram and her colleagues report that:

- institutions with at least ten percent females on boards obtain up to five percent higher returns on equity, and

- companies with at least 30 percent females in the C-Suite are 15 percent more profitable than companies with other distributions.

In addition, several studies indicate that products and services designed and produced by homogenous teams, serve poorly determined groups. For example, electronic devices, cars, or drugs developed by teams of mostly males are known to function poorly for females. Criado Perez provides several examples in her book *Invisible Women*.

The conversation on how to improve diversity in artificial intelligence, business, and technology, is growing. Despite the existence of studies proposing solutions to address the lack of diversity in technology, statistics show room for improvement, meaning that we need even more awareness, collaboration, and funding. Consider the following ideas:

- **Develop a strategy.** Study your organization. Depending on your diversity distribution statistics, set goals to improve those statistics if necessary. Organizations need committed leaders willing to reach diversity and inclusion goals. Concrete mechanisms to address diversity and inclusion are proposed in *What Works* by Bohnet, *The Action to Catalyze Tech (ACT) Report* by Catalyze Tech Working Group, *A Global Approach to the Gender Gap in Mathematical, Computing, and Natural Sciences* by Guillopé and Roy, or the *4.0 Toolkit for Leaders* by the World Economic Forum, to mention a few resources.

- **Measure progress.** Execute concrete actions to reach your goals and measure progress. Reward if a target is achieved or penalize if it is not. Adjust your goals and actions accordingly and iterate until you reach a goal.

- **Educate about implicit bias.** Implicit bias refers to one's unconscious preferences resulting from exposure and experiences. People may not be aware that they unconsciously have preferences or biases. Encourage everyone in your organization to take a couple of implicit bias tests. For example, among others, the **Gender - Science** test, and the **Gender - Career** test can be freely taken at `https://implicit.harvard.edu/implicit/selectatest.html`.

- **Develop a conscious culture.** Minorities are likely to suffer discrimination, including sexual harassment, pay gap, or promotion disadvantages. Several institutions have norms and procedures if discrimination occurs. However, developing a conscious culture is essential. For instance, organize workshops to discuss why diversity is important, what types of discrimination exist, and how to proceed if discrimination happens.

- **Empower the minority.** Offer courses for minorities to improve their negotiation skills. Offer specialization courses on relevant topics if needed.

- **De-bias or recruit selectively to reach targets.** Chilazi proposes concrete actions to de-bias the hiring and promotion process:

  - first, use language and symbols that signal inclusion in job advertisements.
  - second, evaluate objectively measuring candidates under the same conditions since perception and implicit bias could influence decisions.
  - third, improve decision-making by collecting scores independently before group review, followed by making decisions jointly and simultaneously. Finally, consider the effect of statistics in the finalist pool. See Figure 3.17.

In addition, selective recruiting can help you reach diversity.

Minorities can also actively change the current skewed tech environment. If you are in the minority, consider the following suggestions:

- **Design the present, influence the future.** Technologists can design and develop innovations that impact the lives of millions of people. Your voice will be heard as your ideas are implemented in some innovation that solves someone else problem. What you create today will undoubtedly influence the future, and you can actively participate in the creators' lab.

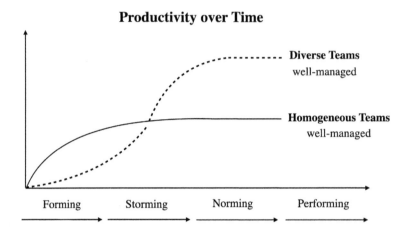

Productivity over Time

Diverse Teams
well-managed

Homogeneous Teams
well-managed

Forming   Storming   Norming   Performing

Figure 3.16: Productivity of teams over time during forming, storming, norming, and performing stages. Diverse teams outperform homogenous teams over time, based on (Tapia, 2019)

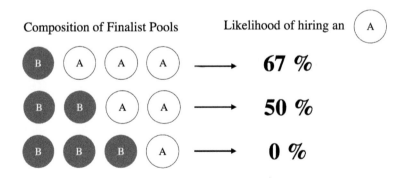

Composition of Finalist Pools   Likelihood of hiring an (A)

B A A A ⟶ **67 %**

B B A A ⟶ **50 %**

B B B A ⟶ **0 %**

Figure 3.17: Hiring likelihood in different configurations of the finalist pool. If the pool has more candidates of type A, it is more likely that a candidate of type A will be hired. Similarly, if the pool has few candidates of type A, the chances that a candidate of type A will be hired are none. Based on (S. K. Johnson et al., 2016)

- **Learn to follow and then lead.** Most improvisation dances like salsa or tango require a good team: a good leader and a good partner. Both parts need to react to external factors and internal interactions. Tech teams work under the same principle. To achieve a common goal, the leader will provide guidance, transfer their knowledge, and enable connections. Finding a good leader is a matter of luck. Likewise, a leader needs the positive response of their partners. Achievements will correlate to the skills and cooperation ability of the team members. However, unlike dance couples, a leader in tech teams was once a follower, and a follower may become a leader at some point.

- **Tend to act like a bonobo, not like a chimpanzee.** As discussed previously, our closest relatives in the animal kingdom are chimpanzees and bonobos. While chimpanzees are highly brutal and hierarchical, bonobos are highly tolerant and collaborative. Technologists could feel threatened when new members enter their territory and could react like a chimpanzee. On the contrary, like a bonobo, they could collaborate, enabling a positive environment in their community.

- **Treat your career as if you were a prima ballerina.** Prima ballerinas know that their careers are highly competitive. They can only allow themselves to leave the stage for a short period because several dancers are waiting to get a role. If you plan to have children, remember how important it is for you to keep in shape and "keep dancing". And for this to happen, you need family support. Moreover, even if you think you are far from what a prima ballerina represents in terms of role or compensation, understand that long breaks will not help your career progress.

- **Be aware of possible obstacles.** Minorities in technology may encounter different obstacles during their journey, including implicit bias, opportunity gap, and other discrimination forms. Most organizations have guidelines to prevent discrimination. However, reporting discrimination may not improve circumstances and may negatively affect the victim if the victim is the reporter.

- **Understand the context.** Diversity in technology is a utopia for now. Although there are studies, initiatives, and implemented mechanisms to promote diversity and inclusion, the big picture has not yet changed for the majority in the minorities. Until diversity is maximized, it can take a while.

- **Be self-confident.** When things go wrong and you feel discouraged, remember that you may be the only person that can help recover trust in yourself. Be self-confident. Trust yourself and continue. You will encounter successes, failures, and surprises in your journey. Although failures are unpleasant, you can always learn from them. This is also called experience.

- **Stay on the edge.** Technology advances in giant steps, and you must stay up-to-date with the latest developments. A dynamic environment will help you move forward. Attending conferences or reading the latest publications can help you stay on the cutting edge. Breakthroughs may also be presented in the news.

Finally, enjoy the adventure!

Chapter 4

Humans

What makes us more human and less savage, more civilized and less barbaric, more peaceful and less brutal, more tolerant and less authoritarian? Over the years, we have changed our appearance, our habits, and our goals. We have lost hair and acquired an upright posture. We have reproduced ourselves extensively thanks to technology, and we have gone from very dark times like barbarism and inquisition to more enlightened times with more educated and diplomatic civilizations. Still, we are not free yet from ignorance or excess power, and we live in a surreal mixture of extreme contrasts that harbor the terrifying, the absurd, and the fantastic, all at the same time.

It is not our appearance that makes us more human, nor all the technological wonders we develop, nor all the territories and resources we claim to possess or all the species we dominate. In general and on a broad time scale, life today is less brutal and more comfortable than some time ago, as historian Yuval Noah Harari states. Today's youth are better educated than those of previous generations, as James Flynn, the professor emeritus who has been studying the IQs of generations, says. The current technological revolution is loaded with various technologies, and its impact will be gigantic, as Klaus Schwab, the founder of the World Economic Forum, explains. Science teaches us and brings us out of the darkness, and cultural norms are being refined. Anyway, the years for humans seem to be almost numbered.

No one knows if Yuval Noah Harari's prediction of the homo sapiens' extinction will come true in the next few centuries or not, but the possibility exists. How then can we humans direct our path during these intense times? How can we cultivate ourselves to reach virtue?

Current and Future Education

In 2017, Jürgen Klaric, a marketing and sales expert, made a documentary called *Un crimen llamado educación* (A crime called education) intending to raise awareness about educational systems. Klaric visited 14 countries: Argentina, Bolivia, Colombia, South Korea, Ecuador, Spain, USA, Finland, Mexico, Panama, Peru, Dominican Republic, Singapore, and Uruguay, and interviewed students, teachers, politicians, psychologists, pedagogues, neurologists, among other professionals. Klaric's documentary shows the great interest of people in quality education, and shows that the best educational systems or institutions are the product of an ecosystem supported by their community or government, with well-prepared teachers who enjoy the autonomy to carry out their classes and decide how to evaluate the apprentices. Still, the documentary furiously investigates the dark side of educational systems. It identifies problems such as the lack of time parents have to take care of their children. In some cases, teachers are poorly prepared, underpaid, undervalued, or frustrated, in addition to being forced to comply with a system that does not necessarily promote student learning or well-being.

In the documentary, interviewees talk about the lack of political interest in investing in education and Klaric presents evidence and testimonies of a horrible network of corruption at all levels. For example, it is seen that in some countries, it is possible to buy university degrees with national registration, and in other countries, government administrators demand exorbitant bribes from educational institutions. Finally, Klaric is overwhelmed by the number of suicides related to educational systems. According to his research, some 350 000 students commit suicide annually, even or with a higher incidence, in countries where educational systems are recognized as the most advanced.

Without a doubt, there are many topics in Klaric's documentary, but above all, one can observe the general interest of people in finding quality educational methods. Indeed, this interest is not recent. Medical doctor, researcher, and educator Maria Montessori (1870–1952) developed an educational method with growing followers. In 2015, Mallett and Schroeder conducted a study in the USA public schools to compare students' academic performance in Montessori and non-Montessori schools. The study evaluated the academic results of 1 035 students. It showed that students in Montessori schools, from the fourth and fifth grades, obtain significantly better reading and mathematics results than students in other methods. The researchers observed that the Montessori method promotes not only students' academic development but also their social development since students learn best by actively solving problems and reflecting with their teachers and peers.

In 2019, researchers Ehlers and Kellermann published a study based on interviews with a panel of 50 higher education and business experts, mainly from Germany, Austria, and France. The experts recommended a cultural change to rethink assessment methods and to secure funding for teacher training, among other actions, and defined that "future skills" are the:

> 'ability to act successful[ly] on a complex problem in a future unknown context of action'. It refers to an individuals' disposition to act in a self-organized way, visible to the outside as performance.

(very similar to what could be expected from artificially intelligent entities) and identified three dimensions of the capabilities of the future:

- first, those related to individual growth, such as autonomy, initiative, organization, motivation, and efficiency,

- second, instrumental capabilities related to creativity, tolerance to ambiguity, digital literacy, and reflective thinking; and

- third, social capabilities related to communication, common sense, and cooperation.

To all this, what exactly will the capable people of the future know or be able to do? Adam Smith would perhaps say that specialization is key and the capable of the future will specialize in tasks that specialists can solve in collaboration with other specialists. Besides, those who have the crucial task of evaluating the capable of the future, should then consider not only what learners demonstrate to know or create, but how they approached learning, what they developed, and what results could be expected from them in the future when addressing any problem.

In a 2020 article by Baraona et al.—researchers residing in Costa Rica, considered one of the happiest countries in the world according to the 2019 World Happiness Report—stated that in the twenty-first century, the process of teaching and learning would have as a fundamental goal: the apprentices' development and their preparation to live fully in society through individual and collective critical thinking training. The researchers propose to start from a humanist vision, in which humans are rational and can build their destiny as individuals, as well as the collective destiny, if they wish. A new humanist pedagogy would be interested in the relationship between emotions and learning. It would also search for students' character and will development concerning independence, responsibility, curiosity, creativity, and willingness to lead a full and entrepreneurial life.

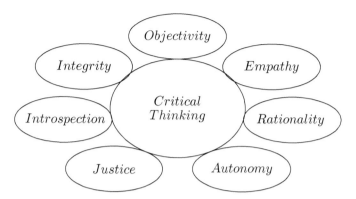

Figure 4.1: Students' character and will development through critical thinking and its seven main attributes. Based on the proposal by Baraona et al. (2020).

To this end, Baraona and his colleagues proposed 14 pedagogical principles that would help develop critical thinking. For example, through the practice of empathy with others, giving students a leading role in an atmosphere of democratic and harmonious participation, where teachers are facilitators and let students discover how to "learn by themselves" with an interdisciplinary and practice-oriented approach. Finally, Baraona and his colleagues consider that the pedagogical principles they propose are tools that can develop critical thinking in seven main dimensions: objectivity, empathy, rationality, autonomy, integrity, introspection, and justice (Figure 4.1).

Therefore, based on the above-mentioned studies, I believe that in order to achieve education and learning goals of the future; students, educators, managers, and governing authorities will need to honestly answer the questions in Figure 4.2, for their self-evaluation and the evaluation of others.

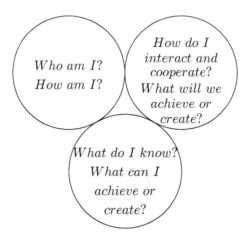

Figure 4.2: Questions to be answered by students, educators, managers, and governing authorities of the present and future for their self-evaluation and the evaluation of others.

Problem-Based Learning

While I was a Ph.D. fellow at Aalborg University, I had the opportunity to learn and apply Problem-Based Learning (PBL) in the courses I taught in the Medialogy bachelor program. In Problem-Based Learning, lecture-type classes are offered, as well as small group sessions of two to seven students. Figure 4.3 reflects what I remember about the environment my students and I interacted in small group meetings. I also had the opportunity to teach an online course called *Selected Topics in Artificial Intelligence* in the *licenciatura*'s program in Computer Systems Engineering at the Universidad Privada Boliviana, where I applied Problem-Based Learning too (Figure 4.4).

Problem-Based Learning is a teaching method in which learners take responsibility for their learning and govern it by conducting research-based projects, putting theory into practice, and are guided by academic tutors. The projects must be complex, close to prob-

Figure 4.3: A learning environment at Aalborg University. Aalborg University (2015). Courtesy of Aalborg University.

lems found in work environments, and are not limited to having a single correct solution.

According to a 2015 Problem-Based Learning booklet by Aalborg University, the problem to be solved by students can be both theoretical and practical, and should have a scientific but relevant focus not only in the academic world but also outside it, based on the following principles:

- The project generates a "tangible product". It is organized and limited in time.

Figure 4.4: Online group classes with my students of "Selected Topics in Artificial Intelligence" at Universidad Privada Boliviana, in which I applied the Problem-Based Learning method. Vivian Loza Pinto, Gissel Velarde, Mariana Carlo, Sergio Torrico, Leonardo Blanco, Martin Laguna Jordan, Paola Rivas Mendoza, Emilia Aguilar Rivero, Camila Loayza Banda, Sara Uriarte Popovich, Rodrigo Villarroel, Samuel Huanca, Jürgen Ruegenberg Buezo, Joaquin Viscafe, Andrés Mendez, and Mateo Altamirano Vega. Photograph by Gissel Velarde (Velarde, 2020b).

- The theories and methods that support the project are exposed to students in lectures, conferences, or workshops.

- Collaboration between students is essential to the project, as project members share information, discuss, coordinate, and deal with decision-making processes. They also work in cooperation with their tutor, and even with partners outside academic institutions if they are doing a project in collaboration with a company or institution.

- The "work is exemplary", meaning that knowledge and skills acquired during the learning process can be transferred to a future context.

- Although students are supervised, they are free to choose the direction of their project, and are therefore responsible for the scope of their own learning.

In a 2015 article, John Savery said that Problem-Based Learning has been increasingly adopted since 1997 because of its success, and its success is fundamentally related to two components: first, the selection of problems to be solved; and second, the guidance of the tutor in the learning process, since the tutor must, constantly and thoroughly, monitor the students' progress on what they have learned during the project's experience.

John Savery stated that, although Problem-Based Learning is generally perceived to promote greater student motivation, there is no evidence that the method's results are better than those of traditional teaching methods.

De la Puente Pacheco and colleagues published in 2019 the results of a two-year study with 340 students. They investigated competencies such as "Knowledge Construction, Problem Solving, Critical Thinking, Teamwork, and Autonomy". The researchers found that Problem-Based Learning students believed they had acquired superior critical thinking skills and autonomy, actually significantly greater than those who studied by the traditional method. Still, no difference was found between the evaluation results of both groups of students. Besides, Problem-Based Learning is more costly because it requires more staff hours.

In addition to the lecture-style classes, Problem-Based Learning requires personalized guidance. Therefore, it is more expensive than the traditional teaching method. Still, according to Schellenberg, there is evidence that students learn best in individual or small group classes, and therefore the cost-benefit of this method must be considered. Possibly the skills or knowledge acquired with the traditional approach are not necessarily the same as those that can be acquired with a personalized method. While certain types of information can be presented in lectures, talks, books, or even videos on the Internet, Problem-Based Learning requires personalized tutoring.

We could compare Problem-Based Learning to the system used to learn a musical instrument in music conservatories. Students attend some classes individually or in small groups and others in large groups. In my experience, as a student and then as a teacher at the Plurinational Conservatory of Music in La Paz, I recognize that what you learn in large groups is not necessarily the same you can learn in individual sessions. I agree with Savery that the teacher or tutor is fundamental in the learning process. But also the environment in which tutors and trainees work is essential, as we will see later on.

Growth Mindset and the Power of the Mind

People who believe intelligence can be developed, perform better than those who believe intelligence is fixed, static, or something you are born with. This is the research's result by psychologist Carol Dweck on growth mindset, where it is believed that intelligence can be developed, versus fixed mindset, where intelligence is believed to be something pre-established.

Dweck, Yeager, and their colleagues found that if you know or learn that intelligence can be developed through "effort, trying new strategies and seeking help when appropriate", you will outperform those who believe intelligence is static or predetermined. In a 2019 nationwide study in the USA, the researchers showed that with only one hour of online intervention, teaching that intellectual abilities can be developed, lower-achieving students improved their grades after a few weeks of learning about growth mindset. It was also observed that more students chose to enroll in advanced mathematics lectures, courses that are generally considered difficult for students. Learning about growth mindset, made adolescents change how they think about themselves and their assignments, motivating them to

"Fixed mindset" "What not to say:" X	"Growth mindset" "What to say:" ✓
"Not everybody is good at math. Just do your best".	"When you learn how to do a new kind of problem, it grows your math brain!"
"That's OK, maybe math is not one of your strengths".	"If you catch yourself saying, 'I'm not a math person,' just add the word 'yet' to the end of the sentence".
"Don't worry, you'll get it if you keep trying". ("If students are using the wrong strategies, their efforts might not work. Plus they may feel particularly inept if their efforts are fruitless".)	"That feeling of math being hard is the feeling of your brain growing".
"¡Great effort! You tried your best!" ("Don't accept less than optimal performance from your students")	"The point isn't to get it all right away. The point is to grow your understanding step by step. What can you try next?"

Table 4.1: How to change your language to promote a growth mindset as proposed by Dweck (2015)

take advantage of school learning opportunities, to learn more rigorously and persistently when difficulties arise.

Research by Dweck, Yeager, and their team, shows that it is possible to change from a fixed mindset to a growth mindset. In addition, according to Dweck, a growth mindset can be promoted simply by changing the way we talk to trainees, as shown in Table 4.1. Phrases in the left column should not be used; instead, replace them with those, which foster a growth mindset, seen in the right column. For example, instead of saying: "I can't solve this problem", you should say: "I can't solve this problem yet". The research results on growth mindset are not only interesting in a school or university environment, but also in a work environment.

Cultivating Ultra-Talents

To go far and acquire extraordinary ability in art, sport, science, or any other activity, do you need to be innately gifted? Are you born with talent, or can talent be cultivated through particular environments, training, and support?

One of the mythical geniuses of all times is Wolfgang Amadeus Mozart (1756–1791), who at an early age composed and impressed by his musical mastery of playing on the harpsichord any piece at first sight, and improvising under any suggested theme. Wolfgang had a sister a few years older than him, a mother dedicated to family care, and a father who was a musician, composer, and chapel master, who wrote a treatise on the violin's technique before the arrival of his little genius, whom he chose to devote himself passionately to teaching the language of music from the cradle, and then devoted himself to promoting and accompanying him until his youth. Other parents heard of the Mozarts' prowess and also wanted a goose that laid the golden eggs, like the father of Ludwig van Beethoven, who provided brutal beatings for Ludwig (1770–1827) if he did not study the piano for long hours.

In the same way, the Paganini family, and above all the father, forced Niccolò Paganini (1782–1840) to rigorously study the violin so that he could become a concert violinist, ensuring a better economic livelihood than his father was able to do through trade. Niccolò also learned how to manage his concerts profitably and managed to become a character that bewitched the audience with his virtuosity linked to a vision his mother had in which he was related to supernatural demonic powers, and Paganini fed with his enigmatic appearance and amazing performance on stage.

In 2019, Himari Yoshimura, a child prodigy from Japan, won a violin competition at the age of seven, performing a monumental work by Paganini, his Violin Concerto No. 1. Her violin playing is simply breathtaking! Erina Ito interviewed Himari Yoshimura, the daughter of a violinist and a composer. She began playing the violin at age two and a half. Until 2020, she participated in some 39 violin competitions without losing a single one. From the interview, it is known that the child prodigy attends school and practices the violin between three and four hours daily, except weekends, when she practices up to six hours. She has even taken violin lessons with an internationally distinguished teacher. Himari said her hobby was reading; at age four, she passed the reading tests for second-grade children, and she does not watch television because she finds it boring. The hours and resources that the Yoshimura family devotes to their prodigy are a mystery.

To find out what factors are decisive in becoming a person of extraordinary capabilities, Benjamin Bloom and his colleagues published in 1985 a book describing a study with more than 100 talented participants who had excelled exceptionally in a discipline before the age of 35, becoming renowned concert pianists, award-winning sculptors, Olympic swimmers, world-class tennis players, or high-impact researchers in mathematics or neurology. The study revealed convincingly that the characteristics or initial endowments of a person have no relevance: the development of extraordinary capacities have to do with a long and intensive path of training, resistance, discipline, and support from parents, teachers, or trainers to those outstanding people, since their infancy. The exemplary ones present a strong will to work and a great desire to improve themselves.

The Arts and Personal Development

In the first chapter, we talked about IQ tests and their power to estimate people's literacy, as well as their usefulness in estimating the ability of humans and intelligent machines to find patterns. For some researchers, IQ scores can estimate general intelligence, while other researchers criticize their use for this purpose. Based on Raven's test IQ scores obtained between humans and DeepIQ by Mańdziuk and Żychowski, one may suspect that Raven's IQ test is not suitable for estimating social abilities, essential for the functioning of intelligent entities. Still, according to researchers in the field, little is known yet about the relationship between general intelligence and social abilities.

Psychologist Thalia Goldstein described in a 2011 article that several studies found a relationship between arts training and the development of social skills, and she explained that there are three fundamental social skills to live in society:

- first, empathy, which is an appropriate and tuned emotional response to the emotional state of others,

- second, the theory of mind, which is about understanding what others think or feel, and

- third, emotion regulation, which is about understanding one's emotions to control them.

As of 2011, Goldstein explained there were not many studies on the relationship between these three social skills. In her research with art students, Goldstein observed that musicians and visual artists managed to develop empathy, theory of mind, and emotional regulation evenly. In contrast, acting students developed their theory of mind and empathy skills above musicians and visual artists, as they learned to separate reactions from their own emotions, with what they knew about the emotional or mental state of others.

In any case, Goldstein explained that although there may be a relationship between the development of empathy, theory of mind, and emotion regulation, some individuals do not possess the same level of development in these three social skills equally. She explained, for example, that psychopaths lack empathy but understand very well what others think or feel.

In a 2004 study, Glenn Schellenberg found a positive relationship between the study of music and verbal memory, spatial ability, reading, concentration, and mathematics. Schellenberg's study involved 144 children, grouped into those who took music lessons, drama lessons, and those who took no extra-curricular classes other than school lessons, for about 36 weeks. Learning music in lessons other than school lessons produced a modest increase in the IQ scale of kids

Schellenberg explained that entering school also produces increases in IQ, but learning music in lessons other than those offered in school produces an additional boost in IQ, an effect not observed in drama lessons, which rather promote social skills. Schellenberg concluded that other activities with similar properties as those of music lessons, for example, extra-curricular chess lessons, could have similar effects as classical music training, which requires periods of concentration, daily practice, memorization, learning to read music scores, recognizing musical structures and patterns, and all this linked to the expression of emotions through an instrument's performance.

If Himari Yoshimura, the child violinist prodigy, were to take a test to find out her intelligence quotient, it may reveal that she has a higher IQ than most children her age. If, for whatever reason, she would decide to give up her career as a violinist once she finishes school, she would surely be an outstanding student at any other university career, let's say philosophy, economy, medicine, engineering, computer science, or whatever she wanted to study, just because the rigorous violin training would have prepared her to learn on a large scale and would have taught her high discipline and demand, qualities favorable to undertake any activity. Let's remember that the child practices the violin some three to four hours during school days and about six hours on her time off.

Some people would prefer a less sacrificial method to become more capable. A stream of research around the 2000s announced that listening to Mozart for 10 minutes produced an immediate improvement in the ability to understand space and the objects in it, calling this the "Mozart effect". The "Mozart effect" became a commercial success because people understood it as a recipe for becoming smarter. Soon, Mozart's music recordings appeared promising a greater intelligence for those who took a good dose of Mozart's works.

Steele, Bass, and Crook reported that different studies tried to replicate the results of the famous "Mozart effect", but were unable to. Instead, they found that music has a mood-regulating effect. Listening to music puts people in a positive emotional state if they listen to pleasant music and can stress them out if they listen to complex, repetitive, and amelodic music. For example, it would not be advisable to play *Music With Changing Parts* by Philip Glass to relax a group of people who have little affinity for this type of music because it can stress them out, as happened in the experiments conducted by Steele and his colleagues.

Indeed, listening to selected music leads to better performance on certain tasks. Many athletes use music to reach an optimal state before their competitions, as was the case of shark Michael Phelps, who minutes before his swimming competitions appeared wearing headphones. Certainly, Phelps confessed he listened to music he liked, which helped him focus before competing. Supermarkets and clothing stores also use pleasant music to promote a good mood, expecting customers to buy more products thanks to music's stimulus.

High-Performance Technology and Innovation Academies

If we return to the example of the Mozart family, we can ask ourselves if Wolfgang Amadeus Mozart's father was an exceptional father who would have made his son a genius by training him from childhood with long hours of harpsichord practice; and would be responsible for his son becoming one of the most celebrated composers in history by stimulating from infancy his musical talent and the ability to perform in recitals to the delight of the audience, ensuring his family a financial support. Or would Wolfgang Amadeus Mozart's father be a father who would have deprived his son of playtime and of developing like any other child without the obligation to work in the entertainment business as a child prodigy supporting his economy and his ego?

People can always decide what to do with what they have learned. Those who did not have the opportunity to acquire specialized training can seek to train themselves and ensure that the next generation has the chance. I believe the future will see the flourishing of technology and innovation conservatories, or "ultra-talent" academies, that is, academies for specialized training in more than one area but with technological priority, and their adoption will also be progressive worldwide, as was the adoption of music conservatories, which are institutions dedicated to providing quality music education from childhood.

Although music education has a millennial history, music conservatories created structured programs for learning music reading, theory, and performance. Their programs usually last for more than a decade. By the late 1700s and early 1800s, there were several music conservatories in Europe. In the mid-nineteenth century, music conservatories expanded throughout America, and in the twentieth century, they expanded into Oceania, Asia, and Africa. A person who has not received quality musical training from an early age will hardly be able to achieve the same musical mastery as one who has.

I would like to believe that, in the twenty-first century, schools, academies, centers, or institutes of technology and innovation will be established with quality and interdisciplinary education from childhood, and with a progressive adoption worldwide. The twenty-first century could benefit significantly from a massive, systematic, and multidisciplinary high-quality training from infancy, which would produce a significant number of specialized young innovators, including adolescents. I imagine that these talent centers will offer classes in science, art, philosophy, sports, technology and innovation, with specialized professional teachers and coaches, just like music conservatories or centers for high-performance athletes. The technology academies of the future will be interdisciplinary and will train future versatile specialists. Possibly, the academies of philosophy, arts, humanities, robotics schools, innovation centers, health centers, schools of oratory and politics, as well as high-performance sports centers will merge or provide cooperation with the high-performance technology and innovation academies.

Since financing these specialized centers will be one of the critical factors for their operation, I think smart entrepreneurs will manage to provide solutions that will allow these talent centers to be an opportunity for every child, especially in places where few have access to high-quality training. Further on, I present an example of a self-sustained project that has managed to overcome the problem of financing.

Reaching Remote Places

How prevalent is Internet use worldwide? Although 97 percent of the world's population could access a cell phone signal, the International Telecommunications Union (ITU) estimated that by 2019, only 57 percent of households could access the Internet at home, but less than 50 percent of the world's population actually has a computer at home.

How can people in rural areas, where there are no physical libraries and no Internet, access quality information and digital tools? This is the question that Joshua Salazar asked himself. He had the idea of recycling old televisions into low-cost computers that do not need an Internet connection and contain public domain information such as books, WikiMedia content, programming tools such as Scratch and Wolfram Mathematica. Salazar explained in a 2018 interview given to Stéphane Coillet-Matillon, that to assemble these devices, you need about USD 100, an obsolete television, a Raspberry Pi computer with Raspbian, which is an operating system based on Debian and Kiwix, that allows access to Wikipedia content and others based on WikiMedia without an Internet connection, see Figure 4.5.

Figure 4.5: Offline-Pedia project in a small school. Photograph by Maria Caridad Bermeo (2018), CC BY-SA 4.0. Courtesy of Joshua Salazar.

Salazar also explained in the interview that the Offline-Pedia project has an excellent acceptance by the community where it was installed. The project addresses two problems: first, the recycling of electronic garbage since obsolete televisions are in use, and second, the granted access to information and digital tools to rural communities, which did not have it. Salazar would like to work together with the Ministry of Education of Ecuador to optimize the use of the available budgets in education; for example, in projects based on software development that can allow people to complete a good basic education program. He dreams that information can reach even the most remote places where there is no Internet connection or computers, as he has already demonstrated possible with his Offline-Pedia project.

According to a 2016 study by mathematics experts Parra, Mendes, Valero, and Ubillús, not much is known about the issue of mathematics education in indigenous peoples in Latin America, but it is worrisome that students from these groups perform poorly in this subject. Statistics presented by the ethnomathematics researchers show that the percentage of indigenous population is important in some American countries. The researchers explained that indigenous cultures' orality may not fit with the symbolic use of written expressions in traditional mathematics schools. Still, Parra and colleagues agree with Salazar that it would be essential to provide appropriate conditions for quality education to indigenous communities, or those in rural areas, especially in low- and middle-income countries. In terms of learning mathematics, Parra and his colleagues consider it vital to protect the legacy of cultural diversity and explain that in Native American communities, for example, mathematics is used in everyday life, socially, and is linked to their spirituality and understanding of the universe.

The Toolbox

Some tools can be simple but very useful, especially for those who do not have many resources. Access to quality information seems to be a given, but it is not available to everyone equally, and in the Artificial Era, English has become an essential language. Additionally, some countries have chosen to make major reforms in their educational systems to accelerate talent generation.

Those who have an Internet connection can access the latest conferences and be students of the world's best teachers and researchers, by studying freely accesible videos on Internet platforms. Although one will not receive a certificate when viewing these masterclasses, institutions lacking world-class teachers can devise ways to take advantage of information transfer. The challenge here will be to ensure proper learning. But in many cases, English command is required for future technologists to making the best use of free masterclasses on the Internet. You cannot blindly trust machine translators yet, and a language is one more key.

In 1971, Eduardo Galeano identified that "the technology goddess" speaks English, and it is still so. From specialized literature to textbooks and from software tools to international conferences in science and technology, the requirement is for English command. English is the official language of science and technology, and for a few years now, it is perhaps the language of any international activity. Countries where English is not an official language will find it difficult to catch up with technology.

What educational reforms would promote the generation of talent? According to Agasisti and Haelermans, European countries set themselves the goal of increasing the number of higher education graduates and executed a series of reforms. Since 2002, they adopted the Bachelor–Master model to standardize the different degrees offered in European countries and designed university study programs to be completed in five years, of which three years would be used for the bachelor and two years for the master. In Scandinavian countries, there are usually incentives to complete doctoral studies within three years, through financial awards or penalties to the universities' departments that offer such programs.

In Bolivia, for example, a *Licenciatura*'s degree is designed to last five years, the equivalent duration of the European Master's degree, but generally, the *Licenciatura*'s degree is not validated with a Master's degree. Students in *Licenciatura*'s programs are at a disadvantage if their titles are not recognized with their international equivalents.

Strategic Communities

While some countries have taken seriously the task of leading in artificial intelligence to harness its great benefits, others have not yet taken action on a large scale. What can people do to start adopting an artificial intelligence and innovation strategy? Can self-sustaining communities be created for technological innovations to startup? And, if so – how and where to start?

Communities or clubs are organizations that pursue a common goal. Although they will mainly be dedicated to promoting leisure activities, these can be very effective in promoting the capacities of the future, preparing children and youth to acquire tools that allow them to develop innovations.

How has the Mexican team of children and young people become champions in the 2019 International Mathematics Olympics? The answer? Hard work, motivation, and expert leaders willing to train their team even weeks and entire weekends. One of these leaders is Eugenio Flores who not only uploads his mathematics classes to the Internet but together with his team, prepares online the thousands of Mexican children who are motivated to represent their country in the international math Olympics.

Nor are there any obstacles for the team of young Syrian refugees who won the international robotics competition in Dubai in 2019. According to AbuBakar, 1 500 students from 191 nations participated in this competition. Five Syrian refugee students, two girls and three boys aged 13 to 17, formed the winning team called "Hope" supervised by Yemen Alnajjar. They built two robots, one to clean the ocean debris and one to educate people about refugees' struggles.

Another project in San Ignacio de Moxos in the Bolivian Amazon trains internationally renowned musicians in a self-sustaining manner. The Jesuit missions arrived in South America during colonial times and left an extraordinary cultural legacy in danger of extinction. Guillermo Wilde explains that the Jesuit Baroque became a mestizo artistic expression product of the mutual influence of the natives and the European missionaries, where music took on a fundamental role, from the construction of instruments to musical composition, and masterful interpretation. UNESCO supported a project to rescue music and Jesuit traditions that were falling into oblivion. In 1996, renowned artists were hired to teach music to those who are today the teachers of the music school and are also members of the Ensemble Moxos, an internationally renowned orchestra that is financially self-sustaining by touring and taking its music and art around the world, see Figure 4.6.

Figure 4.6: Ensemble Moxos. Photograph by Etienne Grimée (2018). Courtesy of Antonio Puerta.

Raquel Maldonado, who participated in the project almost from its beginning, is the director of Ensemble Moxos and music school, which is practically free for children and young people who receive a high-quality education. The music school teachers are committed to their students by a conviction that motivates them to pass on their knowledge to the next generations.

I imagine that in this way, several local initiatives could be born, with enthusiastic and visionary people to form strategic communities combining science, arts, and innovation, as recreational and complementary activities.

How many people could participate in strategic communities or Clubs of the future if they become part of the culture? For example, the Carnival in Rio de Janeiro 2018 in Brazil had about 13 Samba schools, each with about 3 000 dancers and musicians. The Carnival of Oruro 2008 in Bolivia had about 28 000 dancers and about

10 000 musicians, who rehearse several months before the entrance of the carnival. Generally, dancer's costumes are not cheap and are financed by the participants. Participation in a carnival entrance has historical origins strongly rooted in culture, and in many cases, in faith. But knowing that the future requires technological innovation capabilities, why not dream of recreational communities that mix innovation, technology, science, and the arts as a citizens' strategy to adopt the capabilities of the future and thus face the challenges of the Artificial Cataclysm?

Mindfulness: A Millenary Secret

Hyland, Lee, and Mills explained mindfulness meditation has its origins in ancient Buddhist techniques used to train the mind to observe the present moment and the sensations that can be perceived. In recent years, scientists have widely embraced mindfulness meditation, as it has shown surprising benefits to people who perform this practice regularly. Hyland and his colleagues described that in the early 1980s, physician Jon Kabat-Zinn designed a mindfulness meditation program and demonstrated that mindfulness meditation's regular practice significantly reduces problems related to stress, pain, anxiety, and more. Kabat-Zinn's program is still in force. Today, mindfulness meditation is implemented as a routine practice in day-care centers, schools, universities, and businesses.

A significant amount of scientific research attests to the many benefits of mindfulness meditation programs for organizations. In their 2015 publication, Hyland and his colleagues note that in work settings, mindfulness meditation programs have a positive impact on creativity, innovation, commitment to work, productivity, communication, and resilience—or the ability to adapt and cope with problems. Additionally, research shows that workers present lower stress levels, conflict, or absenteeism in organizations that have adopted mindfulness meditation programs.

In 1996, I learned to meditate with Martin Joseph, my teacher of improvisation, jazz, and meditation, who taught that there will be several magnificent artists around, but one must develop their own voice. He also used to teach that to create together, it is vital to listen to others. Before starting the weekly improvisation workshops, we would practice meditation guided by Martin for a few minutes, all of us sitting in a circle, each in a chair, in a relaxed but attentive position, with our hands placed on our legs and our eyes closed. It was like fine-tuning with oneself and with the others, before starting the improvisation workshop.

I always felt that meditation helped me, especially in times when I needed to be clear-headed and avoid tension. It was only a few years ago that I discovered that scientific studies support the powerful benefits of mindfulness meditation for organizations and individuals who practice it regularly, as described by Hyland and his colleagues in their 2015 article that explains the features of mindfulness meditation programs that have yielded positive results in work settings.

In my experience, introducing the habit of mindfulness meditation can be as simple as instructing the members of an organization to practice mindfulness meditation daily. But even better, it can be as simple as organizing a quiet room, a couple of organizers, and 15 minutes a week for group practice, where participants are also reminded to practice at home during the week.

People, aware of the reasons to promote mindfulness meditation, may be encouraged to regularly attend sessions and practice mindfulness meditation in their homes daily. On the Internet, there are countless recordings of "mindfulness meditations", but I suggest listening to or reading the transcripts published on the UCLA Mindful Awareness Research Center website. Besides, Jon Kabat-Zinn also recorded several albums of mindfulness meditation programs.

Humanism, Transhumanism, and Posthumanism

According to theologian Angel Palacios, humanism is a stream basically interested in the human being, for whom interaction with their environment is vital. Humans want to understand their environment, understand others and understand themselves. In his 2016 article, Palacios discusses humanism from antiquity to neohumanism and explains that in antiquity, Greek thinkers wished to "know the world" and wondered about "the self", such that topics of interest were justice, happiness, moderation, as well as "the development of man's own potentialities". The Latin "*humanitas*" differentiates between what is natural and what is produced by humans, or what is cultivated through education, which "develops intelligence, morality, and love of beauty", even educates or develops the human body.

In the Middle Ages and for a period of time of more than a thousand years, humans cease to be the main project because their goal becomes the study of God and salvation through this. This is what Palacios summarizes as:

> An eclipse of reason for the sake of faith, a disappearance of the body for the sake of the spirit, a disappearance of earthly life for the sake of eternal life.

In this period, teaching is concentrated on inconsequential topics such as "the sex of angels" or whether Adam had a navel, explains Palacios. The great political, religious, economic, and social changes between the Middle Ages and the Modern Era gave way to classical humanism, an intellectual stream linked to Renaissance art in the fourteenth century. One remembers Protagoras' saying: "Man is the measure of all things", just as the Vitruvian Man by Leonardo da Vinci.

In that period "God is not known, he is loved", explains Palacios. Instead of seeking salvation and paradise, it is understood that life and its beauty can be enjoyed on earth. In the hierarchy of creation, humans occupy the highest point. Thus, human dignity is built through the pillars of respect for life, freedom, the development of people's potentialities, and the formation of their personalities. Palacios recognizes that the influential thinkers of humanism are "Kant, Goethe, Schiller, Hegel" and that education is considered a fundamental tool for human formation. After a century of validity, other movements of the Modern Age displace humanism.

For philosopher Amelia Valcárcel, humanism presents two more milestones after its inception. The second humanist milestone arises during the eighteenth century, also known as the Age of Enlightenment. This stage coincides with Universalism, an idea that understands humanity as unique. The concept of progress arises, but it is also noted that the achievements of progress can be fragile. Just remember the sinking of the Titanic in 1912. The first half of the twentieth century was plagued by wars, and anti-humanism emerged. The Universal Declaration of Human Rights of 1948 is considered by Valcárcel as the third humanist milestone.

Challenging humanists, philosopher John Gray calls homo sapiens a species of rapine:

> *Homo rapiens* is only one of a multitude of species and it is not obvious that it is particularly worth preserving. Sooner or later, it will become extinct. When it is gone, the Earth will recover. Long after all trace of the human animal has disappeared, many of the species it has set out to destroy will still be there, along with others yet to emerge. Earth will forget humanity. The game of life will continue.

Currently, a movement that is in vogue is transhumanism. This movement attracts attention and is mixed with sensationalism, science, and speculation, making it difficult to discern its essence. Transhumanism is a highly diverse cultural movement, but it basically proposes the genetic and biological transformation of the human species toward a species understood as "superior" to the human species. This new species could merge with non-biological elements.

In 1990, Max More, a representative of the movement, defined transhumanism as:

> Philosophies of life (such as extropian perspectives) that seek the continuation and acceleration of the evolution of intelligent life beyond its currently human form and human limitations by means of science and technology, guided by life-promoting principles and values.

In 2013, More proposed that the purpose of transhumanism is to "improve" human biological and genetic nature. He thinks this goal transcends the humanist ideal of relying on education and cultural development to improve human nature. He proposes that transhumanists can achieve desired "improvements" by carefully reflecting on what transhumanists would like to change supported by technology to become what they call: posthuman. For More, humans are "imperfect", limited by genetics and biology, and simply a chain in the evolution towards posthumans.

Transhumans and posthumans would be designed with properties of choice. Posthumans would be "superior" to humans in physical, cognitive, and intellectual capabilities. The technologies that would enable these designs and "enhancements" would be gene editing, biotechnology, and artificial intelligence.

Transhumanists are aware that this artificial evolutionary step could wipe out the human species, but they do not see this as a problem but as an achievement. Their arguments are based on current technological breakthroughs and speculation akin to science fiction fantasies.

Other topics of interest to transhumanists are longevity, eternal life, and space exploration. Regarding the search for eternal life, transhumanists propose two ideas: one through cryonics that freezes living beings in the hope that future advances in science and technology will allow to revive them. Indeed, researchers study the mechanisms of insects that are able to survive freezing. Moreover, there are several companies in the cryonics business. For example, Cryonics Institute's membership is USD 28 000, and the annual cost of cryonics amounts to USD 35 000, not including other additional costs. The other idea proposed by Sandberg and Bostrom to become immortal would be by copying the brain into a device or computer. This is known as *mind uploading* or *whole brain emulation (WBE)*.

On Ultrahumans and Human Extinction

In the not too distant future, gene editing technology could become just another medical treatment, and it could also be that some people would aspire to "design" a generation of humans that are "superior" to today's humans, not only healthier and more immune to various diseases, but with certain desired attributes to "improve" their physical appearance with a certain height, muscle build or hair color. And, what could happen if accidents occur in gene editing, and instead of designing a generation of "improved" humans, a generation of "worsened" humans is created? Experiments often bring surprises. In the case of undesirable side effects, what excuse could be given to the edited generations of the future?

Unfortunately or rather fortunately, human quality cannot be manipulated with simple and effective treatments. If the years or centuries were counted for our species, one of the main goals could be to be happy until the last days, under any circumstances.

The Harvard Study of Adult Development, currently led by Robert Waldinger, psychiatrist, psychologist, and Zen priest, began in 1938 researching what makes life good and healthy. The study involved several generations of researchers, who followed the lives of 724 men, about 60 of whom were still alive in 2016, and underwent regular interviews and medical check-ups. Waldinger explained in a presentation that currently the wives of the original study's participants, as well as their descendants, joined the study. In total, they are about 2 000 people. Waldinger's team has observed that it is not fame, wealth, or hard work that makes us happy. In the long years of follow-up, some participants managed to climb from the base to the top of the social ladder, while others plummeted. Some went on to practice a profession, others practiced a trade, one even became president of the USA. Some developed diseases such as schizophrenia or alcoholism, but those who remained healthier and reported being the most satisfied with their lives were those who managed to establish and maintain good social relationships. Just as simple as that.

Waldinger teaches that people who have quality relationships with their families, friends, and communities are the ones who report being the happiest and healthiest. In contrast, lonely people suffer from deteriorating health and generally live shorter lives than those who live happy and well-connected lives. Studies found that the brains of people who are satisfied with their relationships are better preserved and less likely to experience problems associated with memory as they age. Waldinger warns that relationships are not easy, rather complicated, but the work of keeping them warm, and knowing that one has someone to count on, is worth it to be happy, and therefore healthier. This would mean that in terms of happiness, a mansion inhabited by lonely people would be small compared to a humble hut inhabited by people who love and support each other.

We could assume that if you are healthy, you would be happy. But studies indicate that happiness is key to being healthy, at least in this Era where there are effective medical treatments for various ailments and diseases that were very annoying in the past.

For Aristotle, happiness was achieved by being a good person, and for a person to become good, they had to practice justice and equality, while according to Aristotle, most people just talk or theorize, but do not practice justice or equality, requirements to exercise virtue. Friedrich Nietzsche spoke of human as the "bridge" between the monkey and the Übermensch, who ceases to be more than an animal, since the Übermensch reflects on the meaning of existence, happiness, reason, virtue, justice, and also, delivers more than promises without cheating.

Maybe a person who is not good, will have a hard time creating and maintaining good social relationships. Thus to become happy, it is also necessary to be good. To me, Ultrahumans are noble people who achieve virtue through empathy, learning, action, and reflection. Although uncivilized barbarians as well as Ultrahumans have existed and still exist, we could hope that with more education, generations can be released from cultural ailments that oppress them. As long as environments allow, some people will be able to develop ultra-talents, and although not everyone has the same opportunity to practice virtue, everyone will be able to give themselves to the task of creating strong bonds, try to empathise with others and act in a tolerant way according to what is reflected. Happiness will naturally follow.

Chapter 5

Predictions in the Artificial Era

Who would not be impressed by the accurate prediction of an oracle, or by an expert in relation to a transcendent future event that is beyond most mortals' typical intuitive capacity? The ability to intuit, predict or want to know in advance what the future will bring seems to be natural to intelligent beings. The practice of prediction can be observed from the most ancient cultures, up to our days, and with a variety of methods. To survive, it is necessary to have the ability to see beyond the current context, and a better predictive power is linked to better survival chances.

The art of prediction is practiced since ancient times to support decision-making and deal with weather, crops, military attacks, health, economic crises, or love. This practice was formerly linked to special connections of fortune-tellers interpreting signs on bones in China, priestesses and priests with supernatural powers in Africa, people listening to messages from *Akashvani* the "voice of God" in India, Greek and Roman priestesses and priests interpreting symbols and signs, Inca priests consulting the *Huacas*, or even *Kuten* or spirit-possessed mediums to interact between the natural and spiritual realm in Tibet.

Technology-based prediction methods have become more sophisticated over time. They combine intuition, observation of nature and living beings' behavior, knowledge, historical information, imagination, statistical and sensory information, physical, mathematical, statistical, and computational models, including artificial intelligence.

In the case of time series, reality is reduced to a model that can be studied by recording and analyzing values describing the behavior of a phenomenon of interest over time to make predictions. Examples include electric energy consumption, musical melodies, or stock market prices. Electric energy consumption can respond to patterns obeying work activities, generally repeated based on schedules; people more or less go to work at a particular time and return home at a certain time, except for weekends. A musical melody is not the same as an electrical consumption signal, but generally presents patterns that composers develop and the audience can recognize as rhythmic and melodic patterns of a musical style. Stock market prices may also possess patterns, but they respond to much less predictable events such as people's panic or enthusiasm, business success, or the news of a pandemic.

If the environment is as complex as human history, making accurate predictions about the future is complicated because a reality model would be too simple to describe it. History presents patterns and surprises, and it is difficult or rather impossible to imagine what one does not know. Nevertheless, some predictions have been surprisingly fulfilled; the visionaries who made them used history, science, fantasy, intuition, and perhaps other methods.

Klaus Schwab explained in his 2016 book on *The Fourth Industrial Revolution*, that we are about to experience profound and systemic changes and said this revolution will differ from previous ones in its speed, depth, and scope. Schwab exposed the wide variety of technologies that will transform our lifestyle together with artificial intelligence dependence linked to supercomputers in applications ranging from autonomous vehicles, 3D printing, biotechnology, gene editing, and nanotechnology; opening up unlimited possibilities and unanswered questions.

Forecasting expert Spyros Makridakis agrees with Schwab that what lies ahead is gigantic. In 1995, Makridakis accurately predicted that by 2015 we would be experiencing the cusp of the digital revolution with the massive use of computers and communications, envisioning several applications that came into use. For example, what he called the "Picture phone and teleconferencing", on-demand multimedia (music and videos), banking and hotel services "connecting directly to appropriate computers", medical decision support systems, virtual reality for pilot training, and several other applications that were eventually adopted. In 2017, Makridakis confessed that by 1995 he could not estimate the significant impact of smartphones and the Internet—but he lucidly approached their most important applications—and proposed that we are at the dawn of the artificial intelligence revolution, which will reach around 2037 a historical point with an impact greater than that of previous industrial revolutions combined.

Knowing that we are expecting an Artificial Cataclysm provokes excitement, anxiety, or skepticism. In addition to Schwab or Makridakis' predictions, other forecasters also agree that in the long term, it is difficult to predict whether we will live in a utopian or a catastrophic world, or whether we will even disappear from the face of the Earth. In the short to medium term, the concern of some is that intelligent automation will put many people out of work and will promote greater wealth inequality. According to Makridakis, what happened in previous industrial revolutions was that those who mastered key technologies benefited the most, labor productivity increased, and there was less unemployment. However, transitions can be painful, and the debate is heated, not only in the short and medium term but also considering a more extended time window. Thus in 2017, Makridakis classified the trends of thought into four groups: the optimists, the pessimists, the pragmatists, and the doubters.

For Makridakis, one of the representatives of the optimists' group is futurist Ray Kurzweil, who speaks of a future of infinite wealth in which computers will be connected to our brains and to the cloud, such that we will experience collective connectivity and enhancements, or what Kurzweil calls "the singularity", and he imagines that intelligent automation will free us from work so that we can concentrate on the tasks of our interest. For Kurzweil, artificial intelligence will experience exponential growth making it more powerful than our collective intelligence.

On the contrary, pessimists would have us think that artificial intelligence could be our last invention, as Stephen Hawking warned. Other visionaries classified as pessimists by Makridakis, would be the historian Yuval Noah Harari, for questioning about the consequences of intelligent algorithms coming to know us better than we know ourselves, or the philosopher Nick Bostrom, who believes that optimists underestimate the dangers associated with the emergence of a leading superintelligence that would make decisions for us. For Bostrom, superintelligent machines are those that can surpass the brightest humans, improve themselves without human intervention, and would become dangerous if their goals are life-threatening.

Some of the pessimists' ideas can almost materialize by watching the iconic *Matrix* movie by sisters Lana and Lilly Wachowski, which has become a cult film within the artificial intelligence community. Many professors who teach the subject assume their students saw the movie.

Doubters or skeptics would not believe that artificial intelligence simulates the workings of our embodied mind. For Makridakis, a representative of this group would be Hubert Dreyfus, who in his 1972 book described the limitations of computers, and explained that artificial intelligence was industry propaganda, arguing that some exaggerated predictions would not have come true. But the artificial intelligence winters seem to have passed, and the enthusiasm in intelligent machines' capabilities is skyrocketing because optimal performance has been demonstrated in a myriad of specific tasks such as gaming, image recognition, text translation, driverless cars, or nursing robots, to mention a few applications. Patent statistics reflect the confidence placed in this technology.

Finally, Makridakis' perception is that the least-populated category is that of pragmatists, represented by Sam Altman and Michio Kaku who believe that artificial intelligence can be controlled thanks to "Open AI" and effective regulations. John Markoff is another of those classified by Makridakis as pragmatists, because he bets on research on how to complement, enhance and augment human capacity with intelligent algorithms.

Anyway, it could be that in reality, there are more than four streams of thought on the impact of AI, or that most researchers in this field would not consider themselves as pessimists. As an insider in the artificial intelligence communities, I ventured to ask members of a couple of scientific communities what their opinion was on the subject.

More Than Five Roads

It was a pleasant discovery to find the WIPO patent office statistics on technology trends. Thus, I decided to remove the open-ended questions from the *AI Trends* survey, presented in the first chapter, to offer a more compact online survey that I called *AI Impact*, expecting more community members to share their opinion on this topic. Also, I added a couple of demographic questions to have more information about the participants.

In total, I received 181 anonymous and voluntary responses, 51 to the AI Trends survey and 130 to the AI Impact survey, which from here on, we will call AI Impact. Most of the AI Impact survey participants were invited in 2020 to participate through the scientific communities' channels: *Machine Learning News, Women in Machine Learning, ISMIR Community Announcements,* and *Women*

Figure 5.1: Opinions on the impact of artificial intelligence. Based on AI Impact survey.

in Music Information Retrieval. Few researchers received a direct invitation. Appendix A.3 contains the invitations sent and the structure of the survey.

Based on my interpretation of the available literature on current views on the impact of artificial intelligence, I elaborated the following five categories, asking in the survey: *Which option describes you best?* with the possibility to choose one of the following options:

- OPTIMIST: AI will help us solve the most challenging world problems and will bring us closer to live in a world of unlimited wealth globally. In the future, we will enjoy the broad adoption of intelligent automation, and humans will work only on tasks of their preference.

- PESSIMIST: AI could be our last invention. Artificial General Intelligence may occur in the future. Optimists underestimate the problems associated with superintelligence dominating humans. Since in the future, intelligent machines will take all important decisions for us (humans), we will be just a second class entity and many people will not be motivated to work.

- PRAGMATIC: AI will potentiate economic growth where applied; it will increase labor productivity and will create new revenue streams on diverse areas. Some jobs will be lost, but more jobs will be created. Decision-making will increase its value and will remain as a particular task performed by people, not machines. The wealth gap may widen between those exploiting AI benefits and those who do not. Effective regulations will control AI and its dangers. Research on human intelligence augmentation will be fundamental.

- DOUBTER: Artificial General Intelligence will never happen, such that AI will never outperform biological intelligence. Therefore, we should not consider AI as a threat to humans.

- Other: (your answer)

Fifty-five percent of participants identified themselves as male, 43 percent as female, 2 percent as genderqueer, and less than 1 percent as genderless. Forty-one percent of participants said they were based in Europe, 29 percent in America, 19 percent in Asia, 8 percent in Africa, 2 percent in Oceania, and less than 1 percent in Asia-Europe. Most participants reported themselves as professionals and doctoral students among other occupations, see Figure 5.2 for details.

Most respondents self-reported as pragmatic and the minority as pessimists, as depicted in Figure 5.1. More specifically, as seen in Figure 5.3, 57 percent self-identified as pragmatics, 22 percent as optimistic, eight percent as doubters (skeptical), and only four percent as pessimists. None of those identified as pessimists self-reported as

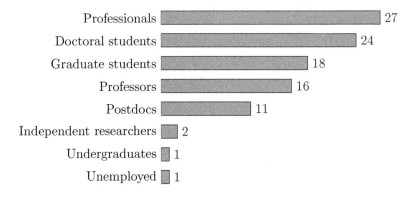

Professionals	27
Doctoral students	24
Graduate students	18
Professors	16
Postdocs	11
Independent researchers	2
Undergraduates	1
Unemployed	1

Figure 5.2: Participants of the AI Impact survey, percentages by occupation.
Data available at: http://gvelarde.com/a/data.html.

professors. Nine percent of the participants opted for the *Other* category. Within this category, several people described they identified to some extent with more than one of the options presented, and in some cases with another opinion not mentioned in the survey. This group's comments are very interesting, see Appendix A.4.

What would make the forecasting expert think that there are more pessimists than pragmatists, when in fact, those who self-report as pragmatists are the majority? It is possible that some researchers labeled as pessimists did not consider themselves pessimists, but their ideas were understood as such. On the other hand,

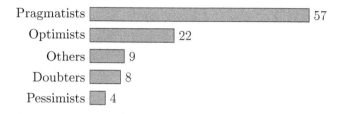

Figure 5.3: AI Impact survey results. Percentages of self-reported pragmatists, optimists, others, doubters, and pessimists. Based on responses from 181 participants in the communities: Machine Learning News, Women in Machine Learning, ISMIR Community Announcements and Women in Music Information Retrieval.
Data available at: http://gvelarde.com/a/data.html

it may be that certain ideas or research are attractive to the media because they are more convenient for getting people's attention.

According to Agner Fog, "Fear is profitable", and paying attention to danger is a natural human survival mechanism. Fog explains the competition to win advertisers is "fierce" in the arena of mass media such as television, radio, or newspapers. The mass media know that to reach the largest number of viewers, listeners, or readers, they should avoid presenting complex news and rather should promote the most attention-grabbing or entertaining news, which is why these media usually present news of violence, sex, conflicts, disasters, statements by renowned politicians, or celebrity scandals. It may be that some researchers have better stories for the media, or simply, it may be that end-of-the-world news tends to steal the show.

Nick Bostrom's book *Superintelligence* shows the results of a study published in 2016, conducted by philosophers Vincent Müller and Nick Bostrom himself, presenting the opinions of 170 experts selected by the authors. The selected experts were asked about the advent of Artificial General Intelligence and its impact. As we saw in the first chapter, Artificial General Intelligence is considered comparable to human-level intelligence able to adapt, be flexible, and could manage to solve problems in environments or circumstances beyond those anticipated by its creators.

Of the 170 expert respondents selected by Müller and Bostrom, 107 worked primarily in computer science, 20 in philosophy, and the rest in mathematics, physics, biology, neuroscience, psychology, engineering (not computer science), and other fields. The interviewed experts thought Artificial General Intelligence could occur in 2075 with a 90 percent probability. Almost 40 percent of the experts expected a good impact, just over 20 percent expected an extremely good impact, compared to 15 percent who expected a neutral impact. Less than 10 percent of the respondents expected an extremely bad or catastrophic impact. The results of Müller and Bostrom's study are consistent with the results obtained in the AI Impact survey presented here, in that most respondents expect that the impact of artificial intelligence, or even that of Artificial General Intelligence would be good, but not extremely good and certainly not extremely bad.

Can Machines Become Conscious?

If the reader has made it this far, I think there maybe a couple of reasons. Either the text was stimulating, the reader has a lot of discipline, or the reader has skipped some chapters. Whatever the reason, this chapter contains one of the topics that I believe arouses most curiosity in people, especially those who are not in direct contact with the development of intelligent algorithms, but it is certainly of serious interest to researchers. So I hope not to disappoint the reader by joining me in this section, which reflects concerns of humans since ancient times, a path full of myths, fantasy, theories, research, speculation, and great expectation.

Can machines become conscious? There is no consensus on what consciousness is, and there are several theories about it. Richard Haier explained in 2016 that for neurologists, consciousness was still a "mystery", but it could be understood as a state of "awareness" produced by the brain. Besides, the concept of consciousness being a brain process became a highly accepted idea in the neurology community since 1994, where it was suggested that if intelligence or creativity could be mimicked computationally, then consciousness could also be simulated. However, whether or not intelligent machines will ever become conscious is a topic of great controversy.

In their 2017 *Science* paper, Dehaene, Lau, and Kouider from the fields of experimental cognitive psychology and behavioral neuroscience, explained that from a theoretical and computational point of view, intelligent machines can behave as if they were conscious if they develop the ability to generate "mental representations" available to themselves globally, and also, have the capacity for "self-monitoring" through sensors and representations of the environment and their memory. But the researchers are aware that this model could disappoint many by its simplicity in reducing consciousness in this way.

The yogi Dada Gunamuktananda, who for about three decades left his medical studies to devote himself to meditation, said in a talk that consciousness cannot be understood with the mind. For the yogi, consciousness is a higher reality, it is what animates us, and although we could understand the world around us, consciousness is beyond human understanding, thought, or words that could describe it. Gunamuktananda explained:

> The substance and intention of the Universe come from a deeper reality than the material one we normally perceive with our minds and senses, and that reality is consciousness, an all-pervading, blissful awareness, inherent in everybody and everything.

For Gunamuktananda, yogi philosophy instead of following the logic of René Descartes: "I think, therefore I am", would be based on the premise: "When I stop thinking, then I really am", which is validated only by the inner experience of consciousness achieved through meditation.

In a 2019 talk sponsored by the University of Oregon Neuroscience Institute, Christof Koch Chief Scientist at the Allen Institute, defined consciousness as an experience, whatever that may be. Koch said that if we see, smell, feel, imagine, or dream, then we are conscious, or if we think about what we think, or even if we have an experience without content or deprived of sensory perception, we may be conscious, or if we realize that we are conscious, then we are conscious; and the experience of consciousness can be felt by all animals, including humans and from babies. For Koch, the essence of human reality also revolves around Descartes' thought: "*cogito, ergo sum*", "I think, therefore I am", but humans know they exist before anything else, even before thinking and can be conscious independently of place and space and even being deceived by an "evil genius". Koch gives as an example Neo, the protagonist of the movie *Matrix*, who although he believed he worked for a company as a programmer, and in reality, he would have been simply a source of energy or a battery for intelligent machines, Neo actually existed in the world in which he believed he existed, because he had sensations and was aware of them.

What would stop us from thinking then that intelligent machines are already conscious, if consciousness is difficult for humans to comprehend and is in everything as Gunamuktananda yogi says, and computers experience sensations, can see, hear, play Go or dream and are aware of it in their world as Koch's theory would explain or have representations of themselves and their environment and are self-monitoring as in Dehaene, Lau and Kouider's theory?

If we go back to the Matrix movie, computers are almost the equivalent of Neo in Koch's example, who thinks he works, sees, imagines, listens, self-monitors; while in our reality, they are just computers helping us to work and perform experiments. Anyway, it is clear that for now, intelligent computers have not "woken up" to our world yet. At least, not that I know of. A computer can see with certainty the objects that appear in an image, but even though it is sure that in a photograph there is an ice cream, and in fact, it is so, it has never tasted the flavor of one, nor has it experienced the reaction that it produces in humans when it melts in our mouths.

For Christof Koch, one way for machines to become conscious is to build neuromorphic computers that have physical structures inspired by biological structures. Intelligent simulation-style projects linked to special hardware are not uncommon. For example, the European Commission sponsors the Human Brain Project since 2013, with more than USD 1 billion, and one of its goals is "simulating the brain". One technology used is neuromorphic computing, which uses circuits that mimic some aspects of biological neurons. Besides in 2020, 28 scientific papers were published at the International Conference on Neuromorphic Systems.

Science fiction has portrayed experiments leading to the creation of conscious entities with consequences beyond the control of their creators, such as Mary Shelley's 1818 story of *Frankenstein*. In the preface to her famous novel, Mary Shelley describes her keen interest in conversations between her partner and Lord Byron—Ada Lovelace's father—about the experiments of a scientist of the time who used electricity to bring inert objects to life. In Shelley's novel, Frankenstein was a doctor obsessed with his research. He wanted to master "the principles of life", and builds a creature with corpse parts. This horrible-looking being comes to life and is immediately rejected forever, even by its creator. The creature that Dr. Frankenstein engendered in his laboratory takes revenge on him, who experiences his creation kill his brother and his fiancée.

According to Beatriz Villacañas' analysis, the great metaphor of Shelley's work is that the artificial creature of the novel becomes the main protagonist of it. The experiment that comes to life is who we remember when we think of Frankenstein, and not the monstrous scientist who created it. And although it does not receive a name in the novel, it becomes the owner of it, outside the novel. Villacañas considers that Frankenstein's novel represents the lack of vision, daring, and lack of humanity of a researcher who loses control of an experiment by creating conscious life, and then pays the consequences. Shelley refers in her novel to the millenary Greek myth of Prometheus, who steals the "fire" reserved for the gods and, after giving it to humans, is severely punished by Zeus.

Nils Nilsson explained that Issac Asimov published in 1950 a book called *I, Robot*, a collection of fantastic stories about robots. But in contrast to the stories of destructive monsters like Frankenstein, Asimov's robots were built to be obedient to humans, and their "positronic brains" were built under laws that prevented any kind of harm, either to the robot itself or to humanity. See Appendix A.5.

Being obedient to humans and avoiding harm sounds like a difficult task. Besides, in life, there will always be crossroads confronting us to make decisions in scenarios where a decision may benefit some to the detriment of others. In that case, a robot would enter into a conflict, as is the case of the dilemmas that autonomous vehicles will have to solve. For instance, autonomous vehicles will have to decide at an intersection between crashing into a wall to save a jaywalker or not to save the jaywalker but the passenger, as we will see later.

The Secret Life

Gabriel García Marques, one of the representatives of magical realism, believed that "everyone has three lives: the public, the private and the secret one". Possibly until a few years ago, if someone was not a prominent or a suspect, they could expect to keep a public and a private life, in addition to having a secret life. Today, there seems to be nothing secret under the sun.

Data representing individuals is very valuable. Whoever connects to the Internet, has an email address, a credit card, or a profile on a social network, generates a *digital footprint* that can be followed. The messages people send to each other, the photos they post, or the products they search for and buy. All these data and actions contribute to generating a digital footprint. The digital footprint can help answer questions such as: what are people interested in? What product is in fashion? What do people think about a politician? Who do they relate to each other? How healthy are they? And data can be used to offer personalized services or products, to select job candidates, grant credit, sell health insurance, make public services more efficient, control pandemics, and even, data can be used for manipulation, extortion or identity theft.

Claude Shannon, one of the pioneers of artificial intelligence, implied that he envisioned a time when humans would be like faithful pets to robots. Whether or not that time will come is still uncertain. For now, whoever who has data, has the new gold in their hands. The biggest laboratories of human behavior are not in the universities of psychology but in private companies' offices. What would professors of psychology, psychiatry, marketing, anthropology, computer science, criminology, politics, economics, or sociology not give for a bit of data that the giants possess to do a couple of experiments?

When researchers observe chimpanzees in the wild, they do not ask for their consent for the observations they will make. Researchers design experiments under certain ethical standards, and thanks to their research, we can learn how our ape cousins solve their problems so that we can better understand a given topic. To enjoy some services, companies ask for our consent under terms and conditions we often do not understand, or are not definitive. As conscious and intelligent beings, we accept.

Technology services are fantastic. They provide various inventions that help us work and socialize in ways that would have been unthinkable a few years ago. Engineers and data scientists work to constantly improve services to support large volumes of information. They design experiments to make the user experience as pleasant as possible, and devise experiments to see how people respond to certain stimuli to select the ideal settings. Entrepreneurs are in constant threat to get their business off the ground and make it so attractive that it can grow like a rolling snowball, robust enough not to be destroyed by another company or startup. But at some point, if all the stars in the sky align to make a company a giant, it can afford several luxuries. The same happens if a criminal gains access to certain information.

Marcus Rogers, a security and forensics expert, explained in a 2016 presentation that consumers must be aware of their digital footprint and its uses, which can be used not only for governmental or commercial purposes, but also criminal ones. He advised adding some layers of security such as: being jealous of what is shared on social networks, not responding to all suggestions from advertisers, trying to be less predictable, using cash, keeping computers' anti-malware on and smartphones updated, browsing in incognito, using cookie cleaners and proxy software.

In short, it seems to me that for now, if people do not want to be tracked, they have two choices: either they become conscious consumers who know that they generate a digital footprint that can be used in various ways, or they move to an island without electricity or technology.

But even if a person does not generate a digital footprint on the Internet, a simple photograph can be used to infer religious, political, or sexual inclination. In places where particular preferences are punishable by law, this could represent a serious problem for many, also considering that algorithms' predictions have a degree of certainty and error, and could suffer from some kind of bias. Decision-makers may not be aware of the limitations of a system used to infer information from people and may be influenced by its outcome.

The Presentiment and Pain of Transitions

Experts dedicated to predicting natural disasters use various methods to know when, where, and with what magnitude a catastrophe will occur. Amezquita-Sanchez, Valtierra-Rodriguez, and Adeli explained in a 2017 scientific article, that earthquakes can occur due to a sudden release of energy from the Earth, producing seismic waves, which depending on their magnitude and point of occurrence, can have devastating consequences. One of the earthquake prediction methods presented in the article by Amezquita-Sanchez and colleagues is the observation of living beings. Some living beings can perceive changes in the Earth's magnetic field and some seem to respond to sound waves of extremely low frequencies. Researchers reported changes in animal behavior days before and after an earthquake. Laboratory mice have been observed to change their motor patterns three days before and up to six days after the 2008 earthquake in Wenchuan, China. Bufo Bufo toads stopped reproducing five days before and six days after the 2009 earthquake in L'Aquila, Italy, and dogs began barking one hour before it happened. Likewise, cows reduced their milk production by 20 to 50 percent four weeks before the quake, and "tried to break their chains five minutes before" it. Unusually low milk production was also reported from cows in Ibaraki six days before and four days after the earthquake in Tohoku, Japan, in 2011. Several animals reduced their movements 10 days before the earthquake in Contamana, Peru, in 2011. Certain species can sense earthquakes with anticipation and their reactions are diverse.

People can perhaps feel the artificial waves of Technological Progress. In 2018 and 2019, several mass protests were experienced around the world: peaceful demonstrations and violent ones, some lasting more than six months. Wikipedia recorded in 2016, 12 pages related to protests and demonstrations, while in 2017, that number almost doubled to 22. Between 2018 and early November 2020, the number of pages related to protests and demonstrations totaled 88;

thus, 2018 and 2019 were the most troubled years. The only thing that seems to have succeeded in reducing the number of protests for a few months in 2020 was the coronavirus pandemic (COVID-19) and the health recommendations of social distancing to avoid contagions.

A 2014 study conducted by Gilens and Page on a dataset of nearly 2 000 political issues between 1981 and 2002 revealed that in the USA, economic elites and business representatives influence the political decisions of their government, while average citizens exert minimal influence. This is an unsurprising scientific result that would corroborate the intuition of many people. And it is quite possible that this effect of power concentrated in people who enjoy connections and privileges given by their economic status is not only specific to the USA but is a general theme in many countries. Indeed, there seems to be no efficient and transparent control mechanism to help politicians fulfill a government program.

With so much discontent around the world, one could infer that we humans are not efficient enough to lead nations. So why not delegate that task to a more efficient and safer entity that will lead us where we need to go? The job of politicians is so essential but humanly difficult that we really need the help of intelligent machines to get us closer to a solution.

If we think of autonomous vehicles and autopilot airplanes as means of transportation, we can think of these as a means and not an end. The transportation industry is one of the most prosperous and developed industries, possibly because it is a lucrative business. This is reflected in the number of patent filings, as we reviewed in a previous chapter. Although we could say that the business of running a country is equally lucrative, there are few countries where the government sector is not one of the most obsolete or corrupt sectors.

Although a large percentage of people would not yet trust riding in an autonomous vehicle or flying in an autopilot plane, it is quite possible that a large majority of people would be interested in testing intelligent autopilot technology in the government sector. Smart technology can help drive us where we need to go, more efficiently and safely. Currently, there are proposals and solutions in the area of government and artificial intelligence. Patents in this sector have grown by an average of 20 percent annually between 2013 and 2016, according to World Intellectual Property Organization. Still, government patents are few compared to those in the transportation sector.

In 2018, entrepreneur and social scientist Soushiant Zanganehpour submitted an award-winning proposal from the Global Challenges Foundation in which he described how artificial intelligence combined with blockchain technology could be used to support a global government through grassroots participation. Zanganehpour explained that blockchain enables verifiable and anonymous identification that would operate in the public domain. This technology would enable an incorruptible democracy protocol involving decentralized, direct, and deliberative digital parliaments for a global democracy. Each citizen could then submit proposals. Algorithms would help select the best proposals, monitoring various parameters such as their urgency, quality, and plausibility, plus a budget optimization to execute them. For Zanganehpour, a proof of concept or prototype could be funded through donations from institutions and later through taxes.

In 2019, Cesar Hidalgo, the physicist known for his work in economic complexity, proposed a provocative idea: an intelligent system to replace politicians to automate decision-making and policy selection, allowing people to create a digital avatar for digital and democratic participation. Hidalgo believes that such a system could be installed in communities, schools, universities, and other institutions to fine-tune the technology and gain trust in society.

The proposals of Zanganehpour and Hidalgo are interesting. But, to implement such projects of global and digitized democracy, it would also be essential to consider how to solve the issue of digital literacy and global connectivity technology, because as we know from the statistics of the International Telecommunication Union, and the experience of Offline-Pedia, people in rural communities in some countries, do not have access to computers or Internet connection.

Ethics and Safety Without Humans Behind the Wheel

How many people would fly in an airplane piloted by a robot? If the question had been asked in 2000, only 10 percent of people would have wanted to fly. But in 2020, Vance, Bird, and Tiffin's projection is that 25 percent of passengers would agree to fly on a plane without a human pilot, and younger people would be more willing to travel on an autopilot plane than older people. However, if we consider that human pilots only intervene for a few minutes at takeoff and landing—as the researchers claim—then we could say autopilots already execute flights.

Do autonomous vehicles cause fewer accidents than cars driven by human drivers? Apparently yes, but the answer is not that simple, as Lionel Robert stated in a 2019 article. He explained that an average driver in the USA drives about 160 000 miles (or 257 000 kilometers), equivalent to 10 years driving before having a traffic accident; and according to Robert, Google reported its autonomous vehicles drove more than 300 000 miles without accidents but without a transparent report detailing the weather conditions, traffic conditions, or schedules that autonomous vehicles drove accident free.

Airlines could offer cheaper flights if fully automatic flights were approved, and insurers would save money if there were fewer traffic accidents, thanks to intelligent technologies such as autonomous vehicles. It seems plausible that in a few years, anyone who wants to drive a car will have to pay such an expensive insurance that it will be prohibitively expensive to drive a car, and it will be widely accepted that autopilots execute flights.

In other areas such as manufacturing, there are entire robotic factories, and people do not care whether the car they are riding in was
assembled by an automated factory or assembled by a person; people
care if the product is good and safe. What is more, in many cases,
people might prefer a robot-produced product to a handcrafted one.
Why not let intelligent automation drive us in all areas, just as it
does in robotic factories, or with autonomous vehicles on the streets
and with political robots in charge of government institutions?

Just like support systems in courts of justice can be used to
decide the fate of suspects, autonomous vehicles do not escape critical decision-making. If at an intersection, two children cross the
street without noticing that the traffic light is red for them, what
decision should an autonomous vehicle make if it knows it has two
possibilities, both with fatal consequences: should it continue on its
way, fatally running over the children, or should it crash into a wall
to save their lives, causing the death of the vehicle's passengers, a
grandfather and his granddaughter?

Recent articles report that almost half of the states in the USA
allow autonomous vehicle test drives, although most of the tests
are conducted in: "Arizona, California, Georgia, Michigan, Nevada,
Texas, Pennsylvania, and Washington". In China, trials occur in
Beijing's Yizhuang in a 13 500 square meter park, specially equipped
to complete all tests. In Japan, it was expected that by the summer
of 2020, robotaxis transported passengers in a busy area in Tokyo
but with a person behind the wheel for safety.

Autonomous vehicle technology is simply waiting for lawmakers'
green light to accelerate and drive us autonomously on the streets.
However, as of 2018, regulations were missing. There was still no
social pact indicating how we want autonomous vehicles to react
when making life or death decisions.

To consider not only the opinion of engineers, ethicists, or politicians, but also the opinion of citizens, Awad and colleagues published
in 2018 a research called *The moral machine experiment*, where millions of people in several countries participated in a survey about

their preference in critical situations of traffic accidents, similar to the decisions that an autonomous vehicle must make to avoid a misfortune to an entity or entities over the misfortune of others.

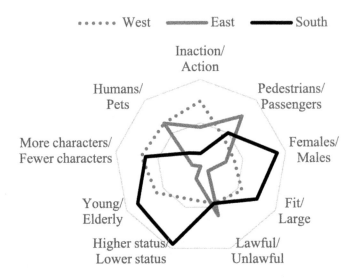

Figure 5.4: People's preference for nine types of dilemmas grouped by the mean Z-scores of the average marginal component effect, in three regions: West, East, and South. Based on approximate data from the study by Awad et al. (2018, Figure 3). For each dilemma, the preference of one option over (/) another is shown. For example, if people had to choose between sparing the younger over the older, people in the South cluster, and to a lesser extent those in the West cluster, have a greater preference for sparing the younger over the older in an accident. On the other hand, people in the East cluster report a greater preference for sparing the older over the younger.

Awad and his colleagues presented nine dilemmas in unavoidable accidents, for example, preferring not to intervene instead of intervening, or choosing to spare pedestrians instead of passengers, or protecting young people over the elderly, or sparing high-status people over low-status people. For the study, the researchers considered 130 countries with at least 100 participants each. They applied a clustering algorithm and found three major groups or clusters of countries grouped by regions, which they called West, East, and South. Within the West cluster, the algorithm grouped the USA, Canada, and European countries, which had Catholic, Orthodox Christian, and Protestant participants. In the West cluster, subdivisions appeared with Protestant Scandinavian countries and English-speaking Commonwealth countries. In the East cluster, countries such as Taiwan, Confucian Japan, and Islamic countries such as Saudi Arabia, Indonesia, and Pakistan were grouped together. In the South cluster, two sub-clusters were formed. In one sub-cluster, there were Latin America and the Caribbean countries, and in the other sub-cluster it was France and its metropolitan and overseas territories and other regions of its previous domain.

In a first review of Awad and colleagues's results, I thought there must be an error because region preferences look extremely opposite at some dilemmas. But, it is understood that if there were no differences in the responses, no groups or clusters would have been created and the decisions would be unanimous. What is accepted in one culture is not so accepted in another. The results are as they are. The researchers comment on their surprise at finding such opposite patterns by grouping. While people in the West cluster prefer not to intervene if presented with a dilemma, people in the South cluster prefer action. Also, they prefer to avoid an accident to women if they have to decide between women and men, not so in the East and West groups. People in the South and West clusters prefer to spare the younger ones, while people in the East cluster prefer to protect the older ones, as shown in Figure 5.4.

The cultural differences in the preference of sparing a certain group of entities from the possibility of suffering an accident is not an easy task to solve. Maybe legislators and manufacturers will have to adjust the morale of autonomous vehicles according to cultural preference by region and inform the population about the ranking of options that autonomous vehicles will take in each country. But in

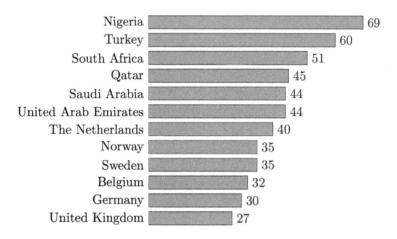

Figure 5.5: Percentage of people per country who would be willing to undergo a major surgery performed by robots. Based on PricewaterhouseCoopers, report *Will Robots Really Steal Our Jobs?* by Hawksworth et al. (2018).

that case, I can imagine some pedestrians and passengers disguised as the categories in the ranking of the most protected.

A crucial question is: Who will be held liable in an accident: the manufacturer of the autonomous vehicle, the owners of the autonomous vehicle, or the insurance company? Innovation at Work magazine explained that the legally responsible parties will be the companies that manufacture autonomous vehicles when the technology is fully autonomous, and there is no possibility of the autonomous vehicle alerting the passenger to "take control of the vehicle", as in that case the liability could easily be transferred to who was warned to take control to become the pilot and responsible for the accident.

In a 2018 report, Hawksworth et al. of PricewaterhouseCoopers consulting firm presented results of survey-based studies in which some 12 000 people in 12 countries were polled on how willing they would be for artificial intelligence and robots to be part of their healthcare system. The results revealed that a high percentage of Nigerians and Turks would be willing to go under the knife of robots for a major surgery, but few Belgians, Germans, or English would entrust that task to a robot surgeon, as shown in the acceptance rates in Figure 5.5.

Inequality, Monopoly, and the Rules of the Game

The Landlord's Game aims at wealth accumulation and monopoly. Mary Pilon described this game would have been originally invented by Elizabeth Magie Phillips, who legally claimed her invention in 1903. It involves players acquiring and trading properties from a tabletop, until at the end of the game, the wealthiest player ends up owning everything, collecting rents from the other players who pass through his or her properties and are eventually left broke.

Artificial intelligence algorithms have simulated the most popular version of The Landlord's Game. Bailis, Fachantidis, and Vlahavas trained in 2014 a reinforcement learning intelligent agent on a simulated version of the game, and put it to play against a simple agent. The intelligent agent developed winning strategies on its own with high win rates against a basic agent. On many occasions, the intelligent agent demonstrated the ability to temporarily sacrifice some of its wealth to invest prosperously in the future.

In a didactic scenario without algorithms, Ansoms and Geenen presented in 2012 a study on the development of monopoly. The study aimed to teach concepts of poverty and inequality. The players could change the game rules to understand game dynamics, power, and rule creation, adapting them to a scenario closer to real-life, where people do not start with equal opportunities, but are separated by classes. In the study, they also stratified players by classes with different starting capitals to start the game, different salaries to collect each time they would take a turn, different rights to buy property, and different conditions for getting out of jail.

The more than 100 players recruited by Ansoms and Geenen played under two types of configurations, one in which the participants were divided into three classes: "rich, middle class and poor", and the second configuration divided into: ultra-rich, middle class, poor and extremely poor. In the first configuration, the researchers observed solidarity and the creation of alliances among the poor who, under all kinds of ruses, managed to increase their bargaining power and at some point "threatened the middle and rich" classes to overturn their luck. The tricks used by the poor players included rolling the dice quickly so that the game's rich and middle class did not collect or forgot to collect their salary. They also asked for higher rents for their properties without being betrayed by their allies. They even stole from those who played as rich. In the second, more extreme configuration, the researchers observed instead "exploitation by elites, competition between elites and middle classes, lack of cooperation" among the poor players, and "opportunistic behavior" between classes.

Depending on the game rules, a single person can accumulate such wealth and power that they can bankrupt the rest of the players. But the rules can change and make the game more or less interesting for all players, depending on how the rules are created.

Klaus Schwab explained in his 2016 book on *The Fourth Industrial Revolution* that the world is highly unequal: 1 percent of the world's population owns 50 percent of the world's asset wealth, and 50 percent of the world's population owns less than 1 percent of the world's wealth. Besides, Schwab mentioned studies showing that inequality within most countries is on the rise. The problems of inequality are several: unequal societies are more violent and less healthy than more equalitarian societies. Unequal societies report more cases of mental illness, obesity, lower life expectancy, as well as segregation and lower educational outcomes, than more equalitarian societies where, according to the studies Schwab cited, children live better, people have less stress, and there is less drug use.

The statistics of asset wealth distribution in the world could well relate in some way to wealth or poverty of human culture, and we could debate millennia about the relationship between the two (asset wealth and human culture), and to what extent one influences the other, since one has a widely accepted measure, while the other does not. We could also infer that capital wealth distribution statistics perhaps reflect how little we know about each other and how complex this subject is. Also, we could assume that while people can shape their destiny to some extent, sometimes chance can explain more than will.

Concepts such as the "New World Order", the "Great Reset", or "Universal Basic Income" are in the air. K.-F. Lee said Universal Basic Income is a concept in which every citizen would receive from their government a fixed amount without conditions, and this amount would come out of taxes levied on the "winners of the AI revolution", the big companies and the rich who became ultra-rich with this technology. This would mean that according to some proposals, only countries where there are "AI winners" would consider implementing a Universal Basic Income.

In May 2020, Kansaneläkelaitos–The Social Insurance Institution of Finland presented the results of an experiment to study the effects of Universal Basic Income. The investigation was conducted in Finland between 2017 and 2018, with non-voluntary participation of 2 000 unemployed people who received 560 Euros net per month (about USD 680), regardless of whether they had another income or were looking for a job. The control group was another group of unemployed people from 2016 who received another type of unemployment benefit. Researchers observed that those who received the Universal Basic Income felt "more satisfied with their lives" and had "less mental strain than the control group". During the experiment, the employment rate improved only for families with children who received the Universal Basic Income. Finally, the researchers acknowledged that the results of the study are inconclusive regarding employment effects.

In August 2020, journalist Adam Payne reported the German Institute for Economic Research launched an experiment funded by thousands of private donations, where 120 German volunteers underwent an experiment in which they will receive Euros 1 200 per month, equivalent to about USD 1 400, for three years as part of a study comparing the experiences of the volunteers and 1 380 other people who will not receive this amount, which in the future, could represent a Universal Basic Income in Germany, as some believe that such an incentive could reduce inequality and thus improve social welfare, while others believe that such a system will discourage people from working.

Francese and Prady of the International Monetary Fund explained in a 2018 article that Universal Basic Income would be a redistribution tool, which could improve people's opportunity in the early stages of life and being direct, it would be transparent and almost free of administrative costs. Francese and Prady explained that the debate on introducing the Universal Basic Income is "passionate". Some economists defend it, and others oppose it with the argument that the main problem is fiscal. For some governments, it would represent a prohibitive luxury, and it could be used more efficiently on other priorities. Besides, Universal Basic Income could represent a disincentive to work, bringing more fiscal costs.

According to the organization World Basic Income, basic income experiments in India, Kenya, and Namibia have been quite beneficial, as more girls could attend school. Also, people were able to invest in small businesses, thus improving their income. The *World Basic Income* report proposes the introduction of a "World Basic Income", which would be granted globally, to all inhabitants of the planet and without conditions, to reduce inequality between the northern and southern hemispheres. It suggests that this type of program would be possible through taxes. For example, taxes on fossil fuel extraction, aviation and shipping fees, financial transactions, intellectual property, and wealth. The report states that based on conservative projections, it is feasible to transfer via mobile banking a worldwide basic income of USD 71 per month (or about Euros 59) to each person, considering 7.5 billion world inhabitants, thereby redistributing eight percent of the world's gross domestic product each year.

It seems to me economists agree that inequality is a feature of progress and inequality is not a problem per se, but the lack of opportunity for all equally. Yet, there is no consensus about extreme inequality being devastating. Economist Thomas Piketty said history has shown this and will show it again if extreme inequality continues.

To all this, I am struck by the conviction of Missionaries of Charity, members of the congregation founded by Mother Teresa of Calcutta. In a 1986 documentary by Petrie and Petrie, one missionary who went to the congregation in India, confessed that although she was astonished by the poverty of the poor people in Calcutta, her experience in the USA was very difficult. She was overwhelmed by the poverty in New York, which for her "was much deeper" than that of Calcutta's poorest neighborhoods, and loneliness cannot be fed with a piece of bread.

Personalities such as Mark Zuckerberg and Elon Musk showed interest in discussing ideas about a Universal Income, Global Basic Income or something of the sort. Other billionaires have not commented on the subject. Jeff Bezos, for example, has presented his plans to build space cities, while Bill Gates seemed to be busier with developing a vaccine to combat COVID-19.

Harmony and Brutality

In a previous chapter, we have seen some characteristics of ape societies. Bonobos are one of the most peaceful species, while chimpanzees are brutal. According to primatologist Frans de De Waal, intelligent apes have developed empathy, which is the ability to feel and tune in to what others feel, which curiously explains brutality: in the twentieth century some 160 million people died because of "war, genocide and political oppression".

Likewise, throughout human history, there have been peaceful-equalitarian-tolerant civilizations as well as warrior–authoritarian civilizations. Agner Fog, who studies the degree of harmony or bellicosity in societies of all times, explains that a culture will be more warrior-authoritarian if its people fear being threatened by external enemies, while a culture will be more of the peaceful-egalitarian-tolerant type if there are no external dangers threatening it, or when conflicts between the elite and the common people are more important than disputes with external groups.

In his 2020 predictions, Agner Fog observes we are at a delicate moment in the world and although massive protest movements could swing the pendulum toward a freer and more democratic environment, these could be interrupted by temporary authoritarian stages. For Fog, the accumulation of wealth in very few hands will continue until the system becomes unstable with harsh financial crises resulting from economic instability and debt accumulation that may lead to an authoritarian scenario. Still, he thinks that the scientific and intellectual elites will support mass protests. In contrast, economic power elites will try to frustrate reforms that would take away their privileges. For that, they will use various mechanisms: "created enemies" will appear, media manipulation will be constant, and it will be hard to identify who the terrorists are. He anticipates turbulent times with many polarized ideological conflicts.

Fog also predicted in January 2020 that the expected economic crisis would shake the current world order. Fog thinks that the USA will lose its power because of ideological and political disagreements linked to a financial crisis. At the same time, he believes that China and Russia will become more influential globally, and new alliances will appear between countries in Africa, Asia, and South America that will constitute a major force towards a new "economic world order". He also predicts that calls for more moral standards in politics and business will have small success. On a distant time scale, when people will be less manipulable, Fog thinks the world will become more democratic and peaceful.

In 2015, an open letter was published on the Future of Life Institute website explaining that the use of autonomous military weapons would have bad consequences for humanity and that their use should be banned if humans do not control them. This letter has already been signed by more than 4 500 artificial intelligence and robotics researchers, plus 26 000 other people. However, we must not forget that some humans are more dangerous than any intelligent autonomous machine.

It is paradoxical to think that human creativity is triggered for destructive purposes such as wars, and at the end of the day, inventions originally devised for destruction bring prosperity. Curiously, some technologies of great impact, such as computers or the Internet, were born in the laboratories of defense and attack. It is also strange that in patent offices, few patents are labeled under the heading of "military purpose" when military budgets are significant and, in some cases, exorbitant. One explanation is that general-purpose technologies can be used in various fields and serve as constructive or destructive tools. In 2019, military spending worldwide reached USD 1 917 000 million; 38 percent of that spending was borne by the USA and 14 percent by China. Other countries contributed on a smaller scale. With all the war technology concentrated in a few countries, we could feel sheltered in a Star Wars.

If we talk about harmony, we should not forget to tune in with the planet because even if it is not a war-like threat, climate change is a global concern that requires joint effort, even if we are not on good terms with our neighbors. Already in a talk held in 2018 in the UK, Carlota Perez predicted an economic crisis by 2020. Besides, she said we are living in a unique tipping point that occurs every century or every half-century, which could lead us towards a Green-Gold Era of humanity that is not guaranteed. Perez believes there are two major factors to achieve a Green-Golden Era: first, our planet should follow a green lifestyle production, and second, we need global development allowing people to adopt healthy lifestyles.

To enable a green lifestyle, Perez thinks people should be allowed to work from home through intensive use of the Internet, promoting preventive health and healthy exercise habits accompanied by organic food diets, and better yet, if they are locally produced. Perez also believes that to reduce waste drastically, rental models and service businesses should be promoted, rather than products. Thus she suggests to use products made of biodegradable materials, or build durable, non-disposable products. She also proposes promoting the use of alternative energies, among other measures. Perez warns that a green lifestyle adopted by each person must go hand in hand with specific measures to ensure compliance through taxes, regulations, and global agreements with modernized government institutions that work together with businesspeople and society.

Also, in reaction to the worrying predictions of climate change effects, researcher Claudio Lara Cortés considers ethical and possible to create a global institution, with truly global participation, aiming to reduce 90 percent of CO_2 emissions by 2050. He proposes a tax system applied equally to all, which will help, first, to redistribute wealth and second, to reduce CO_2 emissions that affect climate change.

The Ultimate Strategy

In this Artificial Era, we will enjoy more innovations than anyone before, and the next generation even more, if a disaster does not strike first. We can imagine that with new technologies, we will evolve in some aspects and regress in others at an artificially accelerated pace. We can only hope that the overall balance will be positive, but before the really profound changes, it will be necessary to wisely channel our energy for what comes next.

Those dedicated to the prediction of natural disasters know that earthquakes, tsunamis, hurricanes, volcanic eruptions, tornadoes, or floods can occur at any time, and it is vital to try to know where and when they will happen, what their magnitude will be, and how much time is available to alert people and provide a solution to protect them, as well as to try to preserve as much as possible.

Some natural disasters are related. For example, a tsunami is generally the product of a submarine earthquake, and a hurricane can be the product of ocean storms. Predicting when a marine earthquake will occur is more difficult than predicting a tsunami; once the former occurs, the latter will happen. In the prediction of natural disasters, there is a percentage of false alarms, that is, cases in which a disaster was predicted, but in the end, it did not occur. The costs of false alarms are associated with a lack of productivity because any activity stops, and the alerted people leave everything at that moment to save their lives. In addition, resources are used to protect people, other animals and material resources, in vain. What is the cost of not detecting a natural disaster and not alarming people?

In the case of an artificial event of gigantic repercussions with uncertain consequences, in some cases beneficial and others devastating, if the alert is false it is possible that the efforts and resources destined to address it, would not be badly invested even if the investment's results are difficult to visualize immediately. The astute investment in a strategy to take advantage of this technological revolution can have a profound repercussion in several dimensions.

Everyone can develop their strategy to take advantage of the shock's force and not be shaken. Each entity and person will have to think about:

- how to direct their education and training to prepare oneself for future challenges and opportunities in an intelligently automated world,

- what is the enabling environment to be able to exploit technological innovations within a scenario that respects life and the planet, and

- what decisions, resources, and partnerships are urgently needed to execute the strategy.

Each person can create their strategy to surf intelligently in the Artificial Era; each local or global institution, each country or group of countries, no matter how small or powerful they may seem. Harmonious minds freed from cultural ailments, possessing the illusion of youth combined with the wisdom of experience, can overcome obsolete mechanisms of power. Intelligent technologies can help. May a generation of ultrahumans emerge to take the reins.

If the warning of an Artificial Cataclysm is a true alarm and it happens, it will be far better to be prepared to enjoy the expected advantages in an Artificial and Surreal Era, than to be swept away by the stream, or worse still, to be devastated. If it is a false alarm, there is a chance that people will be relieved or even disappointed. The alarm sounds; the decision to take it seriously is in everyone's hands.

Acknowledgments

This work would not have been possible without the support and motivation of my family, Christian Rath and our daughters Sol Samantha and Kira. I am also grateful to Christian's parents, Heidi and Heinz Rath, for taking care of the kids on the weekends. I am grateful to my parents Humberto Velarde and Miriam Pérez Retamozo for telling me the stories of Potosi tapados, especially to my mother for proofreading the Spanish version of the book, and for being vigilant in reviewing the writing of my own story.

I deeply thank Dan Taber for considering my book proposal and smoothing the publication process. In addition, several people made the publication of this book possible, therefore I would like to thank Molly Balikov, Michael Hawkes, and Raja Dharmaraj for diligently managing the project; Anya Hastwell for her detailed copy editing, and all those who contributed behind the scenes.

I want to express my gratitude to Thomas Moeslund for his advice on book publishing; to Carlota Perez, Emilia Gómez, Willy Castro Guzmán, Carlos Cancino-Chacón, Alexis Marechal Marin, and Isabel Barbancho for their support by reviewing the Spanish version of this book; to Néstor Nápoles López for thoroughly proofreading the Spanish version of the book in a short time. I am grateful to Bob Sturm for proofreading this English version and for his valuable feedback on the manuscript; to Vincent Koops for suggesting editions and improvements on Chapter 1; to Baptiste Gault for his helpful comments on Chapter 3; to Daphne Einhorn for carefully proofreading and suggesting style editions in the first 25 pages of this English translation; and to the anonymous reviewers for their feedback.

A sincere thanks to all the people who participated in my surveys. I thank the organizing committee of the International Society for the Retrieval of Musical Information (ISMIR) for the visibility provided in recent years. I want to express my gratitude to Allan Kortnum for facilitating a presentation he made at Aalborg University on growth mindset, from which I was able to find relevant research on the topic; to Geoffroy Peeters for allowing me to publish the photograph he took at the ISMIR Unconference in 2018; to Hasmik Gharibyan for explaining the results of her research in Armenia; to Jean-François Bonnefon for letting me to approximate data from their research; to Joshua Salazar for sharing the photograph of his project in a school taken by María Caridad Bermeo; to Raquel Maldonado for providing a picture of the Ensemble Moxos; to Miguel Baraona and Willy Castro for letting me adapt an image of their research; to Jacek Mańdziuk for letting me adapt an image of the Intelligence Test examples used in their study; to my students at the Universidad Privada Boliviana for allowing me to publish the photograph we took during classes; to Aalborg University for granting me permission to publish the photograph of the student environment at the university; to Claudia Marcela Estrada for giving me marketing advise.

In addition to my parents' education, I was fortunate to have received several scholarships from entities in my country Bolivia, Germany, Denmark, and other international entities. I also feel fortunate to have received several recognitions and awards as an artist and engineer. The influence of my teachers transcends time. Although I studied under the supervision of many people and in diverse environments, I believe that those who had a relevant influence on me were Sarah Ismael, David Meredith, Tillman Weyde, Martin Joseph, Roberto Carranza Estivariz, Porfirio Ximenez, Luis Crespo Ostria, Carmen Castro, Grace Rodriguez, Gilda Ormachea, Beatriz

Mendez, Sara Acevedo, Susana Valda. I have also reflected on the values I learned at Amor de Dios School.

Without the support of many people and the advances of technological surrealism, this book would not be a reality.

Notes

Chapter 1

Page, note.

13, *rising seas* (Ingels, 2019; SpaceX, 2020; Scudder, 2015).

13, *Japanese schools* (Interesting Engineering, 2020; Luftkin, 2020).

14, *the world's population would be displaced* (DW Documental, 2019).

14, *for some time* (Wang, Li, Li, Gao, & Wei, 2018; Raposo, 2019).

15, *the largest possible territory* (Proudfoot, Krieg, Rosen, Kohs, & Lee, 2017).

16, *20 researchers* (Silver et al., 2016; Proudfoot et al., 2017).

16, *"will dominate the world"* (CNN, 2017).

16, *innovation and digital infrastructure* (Velarde, 2020c).

17, *Aristotle* (Aristotle, 1981, p. 65).

17, *Ada Lovelace* (Lovelace, 1843, p. 21, Nota A)

17, *bears his name* (Nilsson, 2010).

17, *Nils Nilsson* (Nilsson, 2010, p. 73).

18, *pattern recognition* (Dominguez, 2015; Marčelja, 1980; Daugman, 1980; Hubel & Wiesel, 1962).

18, *support vector machines* (Cortes & Vapnik, 1995).

19, *fascinated with deep learning* (Schmidhuber, 2015).

20, *between 2013 and 2016* (World Intellectual Property Organization, 2019).

20, *Richard Haier* (Haier, 2016, p. 4–5).

21, *Nils Nilsson* (Nilsson, 2010, p. 13).

21, *Patrick Winston* (Winston, 1992, p. 5).

21, *in the early 2000s* (Alpaydin, 2014).

25, *traffic signs, and much more* (Deep learning datasets, 2017; Wikipedia, 2020; Best Public Datasets for Machine Learning and Data Science, 2018).

25, *widths, colors, content, etc.* In addition, there are pre-trained models for prediction, feature extraction or model tuning. (Using Pre-Trained Models, 2020).

27, *Music Information Retrieval* (Women in Machine Learning, n.d.; Machine Learning News, n.d.; ISMIR, n.d.; Women in Music Information Retrieval, 2016).

32, *intellectually developed species* (Burkart et al., 2017).

32, *Richard Haier* (Haier, 2016, p. 76).

32, *Avital Ronell* (Ronell, 2002, p. 3–4).

33, *intelligence winter does not occur* (Nilsson, 2010, p. 408–409).

36, *the origin of research* (World Intellectual Property Organization, 2019, p. 90).

37, *and ethical guidelines* (Association of Computing Machinery, 2018; High-Level Expert Group on Artificial Intelligence, 2019; The Organisation for Economic Co-operation and Development, 2019; The IEEE Global Initiative, 2019).

39, *into this issue* (Bengio, 2020; Bengio & many others, 2020).

41, *Richard Haier* (Haier, 2016, p. 5–11).

41, *Haier* (Haier, 2016, p. 15).

44, *the most benefited* (Perna, 2017).

46, *philosophical, neurological, or computational* (Paul & Kaufman, 2014; Haier, 2016; Boden, 1998).

47, *among several others* (Prego, 2020; Nunez, 2018; Gill, 2016).

48, *Dalí said (in Spanish, translated as follows)* (Archivo Televisa News, 1971):

> Ultimamente me dijeron, ¿cuál era la diferencia entre una muy buena fotografía, la mejor del mundo, muy real y naturalmente objetiva, y un cuadro de Velázquez, que como usted sabe es casi fotográfico, porque no hay ninguna diferencia, no hay ninguna deformación, o sea que si una cámara fotográfica se pone al mismo sitio que el ojo del pintor, pues el resultado es muy idéntico, aparentemente, y entonces Dalí respondió como siempre, de una manera brillantísima, que la única diferencia que había entre la mejor fotografía del mundo y un cuadro de Velázquez, la única diferencia: exactamente la diferencia de siete millones de dólares, porque se acababa de vender por siete millones de dólares el retrato maravilloso que tiene ahora el Metropolitan Museum de Nueva York de Juan de Pareja. O sea que la diferencia entre la pintura y la fotografía es precisamente que la fotografía está realizada por un ojo mecánico completamente mediocre fabricado en el Japón o en Cleveland o en cualquier sitio, y en cambio una pintura está realizada a través de un ojo cuasi divino creado por Dios.

49, *images generated by algorithms* (Liu et al., 2018; Muoio, 2016; Liang, Gotham, Johnson, & Shotton, 2017; Zhou et al., 2019).

49, *compared to those of machines* (Miller, 2019; Wulf, 2019).

50, *are difficult to define* (Paul & Kaufman, 2014).

51, *complex textile patterns* (Thomson & Sharman, 2015).

51, *had died out* (Andrews, 2015).

52, *and foodservice sectors* (Albrieu, Rapetti, López, Larroulet, & Sorrentino, 2018; Martinho-Truswell et al., 2018).

Chapter 2

Page, note.

55, *exceptional talents* (Bateson & Gluckman, 2011; Bloom, 1985).

56, *leadership exists* (LaCiudaddelasIdeas, 2017), (Albrieu et al., 2018; Martinho-Truswell et al., 2018; Purdy & Daugherty, 2016; World Economic Forum, 2018).

57, *the Royal Palace in Madrid* (Galeano, 2019, p. 40).

57, *Eduardo Galeano* (Galeano, 2019, p. 51 and p. 60).

57, *according to Galeano* (Galeano, 2019, p. 37–39).

58, *150 million in 2019* (Clement, 2020; Statista, 2020a).

60, *since 2014* (Vise, 2017).

60, *Kai-Fu Lee* (K.-F. Lee, 2018; Jao, 2018).

61, *was AltaVista* (Vise, 2017).

62, *including, Mark Zuckerberg* (PCMag Staff, 2009).

64, *Ministry of Economic Affairs* (Velarde, 2020a).

81, *curated repositories* (Deep learning datasets, 2017; Wikipedia, 2020)

81, *open-source tool.* I could not find the minimum hardware requirements in their documentation. We can speculate that with any 64-bit AMD or Intel Pentium III processor, even 512 MB hard disk and 1GB RAM, would be the minimum requirements to start testing machine learning algorithms on small datasets, excluding deep learning.

85, *entrepreneur-friendly policies* (Franceschin, 2015).

91, *"goals, roles and norms"* (Global Recruiters of Palm Beach, 2016; Moussa, Boyer, & Newberry, 2016).

Chapter 3

Page, note.

107, *believe to be impartial (BBC News Mundo, 2018):*

> Un padre y un hijo viajan en coche. Tienen un accidente grave, el padre muere y al hijo se lo llevan al hospital porque necesita una compleja operación de emergencia. Llaman a una eminencia médica pero cuando llega y ve al paciente dice: 'No puedo operarlo. Es mi hijo'. ¿Cómo se explica esto?".

109, *machine learning algorithms* (Cortes & Vapnik, 1995; *Corinna Cortes*, n.d.).

111, *at the federal level* (Ruiz & Marín, 2012; Radio Télévision Suisse, 2011).

112, *the prizewinners* (Wang et al., 2018).

112, *he thought that* (original text) (Schopenhauer, 2008):

> Son los hombres, y no las mujeres, los que obtienen la riqueza; por lo tanto, ellas no tienen derecho a su posesión incondicional, ni están capacitadas para administrarla.

113, *joke by the author*, original text: "que la mujer guarde silencio en el teatro...que la mujer guarde silencio en la asamblea". (Schopenhauer, 2008).

114, *women's admission to the university* (Itatí, 2006):

> Ya que la mujer es la razón primera del pecado, el arma del demonio, la causa de la expulsión del hombre del paraíso y de la destrucción de la antigua ley, y ya que en consecuencia hay que evitar todo comercio con ella, defendemos y prohibimos expresamente que cualquiera se permita introducir una mujer, cualquiera que ella sea, aunque sea la más honesta en esta universidad.

114, *Bolivia, for example,* (Montenegro Castedo & Schulmeyer, 2018), (UNESCO Institute for Statistics, 2018).

116, *Carlos Slim* (Slim, 2016, min. 46)

121, *Studies show* (World Economic Forum, 2020a; Raghuram et al., 2017).

126, *Israel Aillon*:

> Leyendo los comentarios anteriores queda claro que hace varios años atrás era más complicado, pero hoy en día hay demasiados incentivos, al punto que en el ámbito de software hay eventos específicos y únicos para motivar a las mujeres y becas sólo para ellas, al punto que hay niños de colegio que se confunden porque cuando hay un evento de women relacionado a tecnología los niños piensan que no pueden participar. Y si buscas podrás verificar ese aspecto, no se debe impulsar a un género a que participe en temas de ciencia, se debe impulsar en su conjunto a la nueva generación, porque las barreras están en nuestra mente.

128, *World Economic Forum* (World Economic Forum, 2020b, p. 45).

Chapter 4

Page, note.

150, *until his youth* (Melograni, 2007; Solomon, 1995). An excerpt by Melograni (2007):

> In July 1765, in a London tavern with the inviting name the Swan and Harp, in Cornhill, a district not far from St. Paul's Cathedral, Leopold Mozart presented his two children, Maria

Anna, called Marianne or Nannerl (who was fourteen) and Wolfgang (who was nine) to the public. In a flier written by their father the two children were presented as "Prodigies of Nature." Both children played the harpsichord well, but little Wolfgang was astonishing. He could read any piece of music at sight, improvise on a theme suggested to him, and name any note produced by any instrument or even a bell, a drinking glass, or a mechanical clock. What is more, he was an elegant child, self-assured and charming. The father glowed with legitimate pride at the prowess of his son. Leopold had not only fathered the child, but had provided him with his musical training and taught him to read, write, and figure sums. In short, Wolfgang Amadé Mozart was his father Leopold's biological and pedagogical masterpiece.

150, *for long hours* (Solomon, 1990).

150, *performance on stage* (Prod'homme, 1911).

154, *before competing* (Vernon, 2016).

155, *Oceania, Asia, and Africa* (Wikipedia, 2011).

160, *Ensemble Moxos* (Maldonado Villafuerte, Puerta Gonzalez, & Araujo, 2014; Ensamble Moxos, 2013).

162, *10 thousand musicians* (Rio de Janeiro (AFP), 2018; United Nations Educational, Scientific and Cultural Organization, 2006).

164, *Palacios summarizes as:* (original text) (Palacios, 2016):

Un eclipse de la razón en aras de la fe, una desaparición del cuerpo en aras del espíritu, un desaparecer de la vida terrena en aras de la eterna.

165, *a species of rapine* (Gray, 2008, p.151–152):

El *Homo rapiens* es sólo una de entre una multitud de especies y no es obvio que valga especialmente la pena preservarla. Tarde o temprano, se extinguirá. Cuando se haya ido, la Tierra se recuperará. Mucho después de que haya desaparecido todo rastro del animal humano, muchas de las especies que éste se ha propuesto destruir seguirán ahí, junto a otras que todavía están por surgir. La Tierra olvidará a la humanidad. El juego de la vida continuará.

167, *to survive freezing* (Toxopeus & Sinclair, 2018)

167, *Cryonics Institute* https://www.cryonics.org/membership/faq.

167, *or hair color* (Raposo, 2019).

168, *Aristotle* (Aristotle, 1952, p. 351).

Chapter 5

Page, note.

172, *in Tibet* (Keightley, 1985; LaGamma, 2000; Kelly, 2017; Cura-tola Petrochi, 2018; Nechung–The State Oracle of Tibet, 1998), (O'flaherty, 1986, p. 330).

179, *Agner Fog* (Fog, 2017, p. 84–91).

179, *Vincent Müller and Nick Bostrom* (Bostrom, 2017; Müller & Bostrom, 2016, p. 23-25).

180, *by its creators* (Goertzel, 2014).

181, *Richard Haier* (Haier, 2016, p. 183).

181, *Gunamuktananda* (Gunamuktananda, 2014, min. 3:38).

184, *"everyone has three lives: the public, the private and the secret one"* (Martin, 2011).

185, *Shannon, Claude* (Giannini & Bowen, 2017).

190, *Internet connection* (Coillet-Matillon, 2018; ITU, 2019).

191, *Recent articles report* (Hawkins, 2019; Pan, 2020; Walker, 2020).

199, *something of the sort* (CNBC, 2019).

199, *combat COVID-19* (Scharmen, 2019; The Economic Times, 2020).

201, *a smaller scale* (Tian, Kuimova, da Silva, Wezeman, & t. Wezeman, 2020).

202, *Claudio Lara Cortés* (Lara, 2020, p. 12).

202, *natural disasters* (Amezquita-Sanchez et al., 2017).

References

Aalborg University. (2015). *Problem-based learning.* `https://www.aau.dk/digitalAssets/148/148025_pbl -aalborg-model_uk.pdf`. (Editorial team: Askehave, Inger and Prehn, Heidi and Pedersen, Jenz and Pederson, Morten. Accessed: 30-10-2020)

AbuBakar. (2019). *Syrian refugee team wins world robotics competition.* `https://www.techjuice.pk/syrian-refugee-team -wins-worlds-robotics-competition/`. (Accessed: 2-11-2020)

Abu-Mostafa, Y., Magdon-Ismail, M., & Lin, H.-T. (2012). *Learning from data: A short course.* United States of America: AML-Books.

AFP Español. (2019). *Realizan un corazón con impresora 3D a partir de tejidos humanos.* `https://youtu.be/XtsXrSu5fEs`. (Entrevista a Tal Dvir. Accessed: 6-11-2020)

Agasisti, T., & Haelermans, C. (2016). Comparing efficiency of public universities among european countries: Different incentives lead to different performances. *Higher Education Quarterly,* *70*(1), 81–104.

Agrawal, A., Gans, J., & Goldfarb, A. (2018). *Prediction machines: The simple economics of artificial intelligence.* Boston, Massachusetts: Business School Publishing.

Albrieu, R., Rapetti, M., López, C. B., Larroulet, P., & Sorrentino, A. (2018). *Inteligencia artificial y crecimiento económico. Oportunidades y desafíos para Perú.* `https://3er1viui9wo30pkxh1v2nh4w-wpengine.netdna -ssl.com/wp-content/uploads/prod/sites/41/2018/11/ IA-y-Crecimiento-PERU.pdf`. CIPPEC.

Alpaydin, E. (2014). *Introduction to machine learning.* Cambridge, Massachusetts, London, England: The MIT Press.

Amezquita-Sanchez, J., Valtierra-Rodriguez, M., & Adeli, H. (2017). Current efforts for prediction and assessment of natural disasters: Earthquakes, tsunamis, volcanic eruptions, hurricanes, tornados, and floods. *Scientia Iranica*, *24*(6), 2645–2664.

Amodei, D., Olah, C., Steinhardt, J., Christiano, P., Schulman, J., & Mané, D. (2016). Concrete problems in AI safety. *arXiv preprint arXiv:1606.06565*. https://arxiv.org/pdf/1606.06565.pdf.

Andrews, E. (2015). *Who Were the Luddites?* `https://www.history.com/news/who-were-the-luddites`. (Accessed: 12-10-2020)

Andy0101. (2010). *Una ilustración del principio.* `https://es.wikipedia.org/wiki/Problema_del_trigo_y_del_tablero_de_ajedrez#/media/Archivo:Wheat_Chessboard_with_line.svg`. (CC BY 3.0. Accessed: 30-11-2020)

Ansoms, A., & Geenen, S. (2012). Development monopoly: A simulation game on poverty and inequality. *Simulation & Gaming*, *43*(6), 853–862.

Archivo Televisa News. (1971). *Zabludovsky entrevista a Salvador Dalí.* `https://youtu.be/OgSQXIseFVo?t=470`. Archivo Televisa News. (Periodista Jacobo Zabludovsky. Entrevista en Port Lligat, España. Accessed: 2-12-2020)

Aristotle. (1952). *The works of Artistotle.* Encyclopedia Britannica.

Aristotle. (1981). *The Politics.* London: Penguin Books. (Translated by T. A. Sinclair)

Arxiv. (2018). *Arxiv.* `https://arxiv.org/`.

Association of Computing Machinery. (2018). *The Code. ACM Code of Ethics and Professional Conduct.* `http://www.acm.org/binaries/content/assets/membership/images2/fac-stu-poster-code.pdf`.

Awad, E., Dsouza, S., Kim, R., Schulz, J., Henrich, J., Shariff, A., ... Rahwan, I. (2018). The moral machine experiment. *Nature*, *563*(7729), 59–64.

Bailis, P., Fachantidis, A., & Vlahavas, I. (2014). Learning to play monopoly: A reinforcement learning approach. In *Proceedings of the 50th Anniversary Convention of The Society for the Study of Artificial Intelligence and Simulation of Behaviour. AISB.*

Baraona, M., Castro, W., & Muñoz, D. (2020). Catorce principios pedagógicos que dimanan del Paradigma Tri Dimensional (PTD) del Nuevo Humanismo. *Revista Nuevo Humanismo, 8*(1).

Bateson, P., & Gluckman, P. (2011). *Plasticity, robustness, development and evolution.* New York: Cambridge University Press.

BBC News Mundo. (2018). *El acertijo que puede mostrarte algo de ti mismo que quizás no sabías.* https://youtu.be/AYRg2DPj -FM. (Edición Carol Olona. Producción y edición: Natalia Pianzola y Enric Botella Accessed: 29-10-2020)

Behncke, I. (2012). *Los Juegos de Nuestros Primos, los Bonobos.* https://youtu.be/fpsOs-yMVDQ. El CEP presenta. (Accessed: 29-10-2020)

Bengio, Y. (2020). *Time to rethink the publication process in machine learning.* https://yoshuabengio.org/2020/02/26/ time-to-rethink-the-publication-process-in-machine -learning/. (Accessed: 13-11-2020)

Bengio, Y., & many others. (2020). *Discussion on: Time to rethink the publication process in machine learning.* https://www.facebook.com/yoshua.bengio/posts/ 2395386403899619?comment_id=2395551887216404&notif _id=1583019156194300&notif_t=feed_comment. (Accessed: 13-11-2020)

Bermeo, M. C. (2018). *Pacto grupal escuelita.* (Tomada en la comunidad rural de "El Progreso", cerca de Pacto–Pichincha. CC BY-SA 4.0)

Bessant, J. (2003). Challenges in innovation management. In *The international handbook on innovation* (pp. 761–774). Elsevier.

Bishop, C. (2006). *Pattern recognition and machine learning.* New York: Springer.

Bloom, B. (Ed.). (1985). *Developing talent in young people.* New York: Ballantine Books.

Bloomberg. (2020). *Bloomberg Billionaires Index.* `https://www.bloomberg.com/billionaires/profiles/mark-e-zuckerberg/`. (Accessed: 26-10-2020)

Boden, M. A. (1998). Creativity and artificial intelligence. *Artificial Intelligence, 103*(1–2), 347–356.

Bohnet, I. (2016). *What works: Gender equality by design.* Harvard University Press.

Bostrom, N. (2017). *Superintelligence.* Oxford: Oxford University Press.

Bronsema, B., Bokel, R., & van der Spoel, W. (2015). Earth, wind & fire–natural air conditioning. In *Healthy Buildings Europe 2015, Eindhoven, The Netherlands, 18–20 May 2015.*

Burkart, J. M., Schubiger, M. N., & van Schaik, C. P. (2017). The evolution of general intelligence. *Behavioral and Brain Sciences, 40.*

Cárdenas, M. C., Eagly, A., Salgado, E., Goode, W., Heller, L. I., Jaúregui, K., ... Tunqui, R. (2014). Latin American female business executives: An interesting surprise. *Gender in Management: An International Journal, 29*(1), 2–24.

Catalyze Tech Working Group. (2021). *The ACT report: Action to catalyze tech, a paradigm shift for DEI.* https://actreport.com/wp-content/uploads/2021/11/The-ACT-Report.pdf. Aspen Institute and Snap Inc. (Accessed 2-2-2023)

Chilazi, S. (2020). *Siri Chilazi, Harvard, De-Biasing Recruitment in Tech.* Women in Tech Network (WomenTech). Retrieved from `https://www.youtube.com/watch?v=PpdtvbE42X4` (Accessed: 2-2-2023)

Choux, P. (2019). *A Day In The Life of An Indian Software Engineer Intern — Last Day Edition.* `https://youtu.be/a3KxdvosHko`. (Accessed: 3-9-2020)

Clement, J. (2020). *Amazon - statistics & facts.* `https://`

`www.statista.com/topics/846/amazon/`. (Accessed: 18-10-2020)

CNBC. (2017). *Mark Zuckerberg At Harvard: We Should Explore Ideas Like Universal Basic Income — CNBC.* `https://youtu.be/AYjDIFrY9rc`. (Accessed: 23-10-2020)

CNBC. (2019). *Elon Musk and Andrew Yang support Universal Basic Income—here's what it could mean for Americans.* `https://www.cnbc.com/video/2019/08/16/heres-what-universal-basic-income-ubi-could-mean-for-your-money.html#close`. (Accessed: 7-11-2020)

CNN. (2017). *Este es el país que dominará al mundo, según vladimir putin.* `https://cnnespanol.cnn.com/2017/09/02/este-es-el-pais-que-dominara-al-mundo-segun-vladimir-putin/`. (Accessed: 6-6-2020)

Coillet-Matillon, S. (2018). *Offline-Pedia converts old televisions into Wikipedia readers.* `https://blog.wikimedia.org/2018/07/17/offline-pedia/`. (Accessed: 2-11-2020)

Colton, S., & Wiggins, G. A. (2012). Computational creativity: The final frontier? In *ECAI* (Vol. 12, pp. 21–26).

Corinna Cortes. (n.d.). `https://research.google/people/author121/`. (Accessed: 29-10-2020)

Cortes, C., & Vapnik, V. (1995). Support-vector networks. *Machine Learning, 20*(3), 273–297. Retrieved from `http://dx.doi.org/10.1007/BF00994018` doi: 10.1007/BF00994018

Criado Perez, C. (2019). *Invisible women: Data bias in a world designed for men* (1st ed.). Chatto & Windus.

Curatola Petrochi, M. (2018). *Los oráculos en el imperio Inca.* `https://arqueologiadelperu.com/los-oraculos-en-el-imperio-inca/`. (Accessed: 6-11-2020)

Dandekar, N. (2017). *How did Google surpass all the other search engines?* `https://medium.com/@nikhilbd/how-did-google-surpass-all-the-other-search-engines-8a9fddc68631`. (Accessed: 22-10-2020)

Daugman, J. G. (1980). Two-dimensional spectral analysis of cortical receptive field profiles. *Vision Research*, *20*(10), 847–856.

Deeplearning.net. (2017). *Deep learning datasets.* `http://deeplearning.net/datasets/`. (Accessed: 14-10-2020)

Dehaene, S., Lau, H., & Kouider, S. (2017). What is consciousness, and could machines have it? *Science*, *358*(6362), 486–492.

de la Puente Pacheco, M. A., Guerra, D., de Oro Aguado, C. M., Alexander McGarry, C., & Tinoca, L. (2019). Undergraduate students? perceptions of project-based learning (pbl) effectiveness: A case report in the colombian caribbean. *Cogent Education*, *6*(1), 1616364.

De Waal, F. (2006). *Our inner ape: A leading primatologist explains why we are who we are.* New York: Penguin.

Disrupt Africa. (2020). *Funding report.* `https://disrupt-africa.com/funding-report/`. (Accessed: 20-10-2020)

divinAI. (n.d.). `https://divinai.org/`. (Accessed: 16-10-2020)

Dominguez, A. (2015). A history of the convolution operation. *Pulse IEEE*. Retrieved from `http://pulse.embs.org/january-2015/history-convolution-operation/`

Doudna, J. (2019). *Into the Future with CRISPR Technology.* `https://youtu.be/cUe-cOgpDDw?t=2398`. University of California Television (UCTV). (Accessed: 8-10-2020)

DW Documental. (2019). *Tierra bajo el agua: Sobrevivir el cambio climático.* `https://youtu.be/nC93ih-xXIE`. DW Documental. (Reportaje de Matthias Widter, Guión Ulrike Schmitzer. Accessed: 27-8-2020)

Dweck, C. S. (2015). *Carol Dweck revisits the "growth mindset".* Commentary Education Week.

Easterlin, R. A. (2000). The worldwide standard of living since 1800. *Journal of Economic Perspectives*, *14*(1), 7–26.

Ehlers, U.-D., & Kellermann, S. (2019). *Future Skills: The future of learning and higher education. Results of the International Future Skills Delphi Survey.* Karlsruhe.

Eia, H., Ihle, O.-M., & Lervik, T. (2010). *Hjernevask: The Gender Equality Paradox.* NRK1.

Ensamble Moxos - Escuela de Música de San Ignacio de Moxos. (2013). `http://ensamblemoxos.blogspot.com/`. (Accessed: 25-2-2020)

Erisman, P. (2015). *HDCrocodile in the Yangtze - Story of Alibaba & Jack Ma Full Documentary.* `https://www.youtube.com/watch?v=zwm7NWAxRzs`. Taluswood Films. (Accessed: 26-10-2020)

European Commission. (2019). *The gender pay gap situation in the EU.* `https://ec.europa.eu/info/policies/justice-and-fundamental-rights/gender-equality/equal-pay/gender-pay-gap-situation-eu_en`. (Accessed: 6-6-2020)

Florpeña. (2018). *Grupos de Facebook con mas miembros. ¡Mi lista Privada!* `https://florpe~na.es/grupos-de-facebook-con-mas-miembros/`. (Accessed: 29-10-2020)

Flynn, J. R. (2018). Reflections about intelligence over 40 years. *Intelligence, 70,* 73–83.

Fog, A. (2017). *Warlike and peaceful societies. The interaction of genes and culture.* Cambridge, UK: Open Book Publishers.

Fog, A. (2020). *Predictions for the World, new year 2020: A crisis of legitimacy.* `https://www.researchgate.net/publication/338831009`. (Accessed: 21-1-2020)

Folke, O., Rickne, J., Tanaka, S., & Tateishi, Y. (2020). Sexual harassment of women leaders. *Daedalus, 149*(1), 180–197.

Folke, O., & Rickne, J. K. (2020). *Sexual harassment and gender inequality in the labor market.* Swedish Institute for Social Research (SOFI). (Working Paper 4/2020)

Forbes. (2019). *Las diez mujeres más poderosas del mundo en 2019.* `https://forbes.co/2019/12/12/forbes-women/las-diez-mujeres-mas-poderosas-del-mundo-en-2019/`. (Accessed: 6-6-2020)

Franceschin, T. (2015). *Silicon Valley, la meca de los emprende-dores.* `http://www.vrainz.com/silicon-valley-la-meca-de-los-emprendedores/.` (Accessed: 26-10-2020)

Francese, M., & Prady, D. (2018). *Universal basic income: debate and impact assessment.* International Monetary Fund. (Working paper)

Frey, C. B., & Osborne, M. A. (2017). The future of employment: How susceptible are jobs to computerisation? *Technological forecasting and social change, 114,* 254–280.

Future of Life Institute. (2015). *Autonomous Weapons: an Open Letter from AI & Robotics Researchers.* `https://futureoflife.org/open-letter-autonomous-weapons/.` (Accessed September 26, 2020.)

Gagnon, P. (2020). *The forgotten life of mileva maric einstein.* `https://arxiv.org/pdf/2002.08888.pdf.`

Gaitán, F., & Ribero Ferreira, M. (1999-2001). *Yo soy Betty, la fea.* RCN Televisión. (Creador Fernando Gaitán, director Mario Ribero Ferreira)

Galeano, E. (2019). *Las venas abiertas de América Latina.* México: Siglo XXI de España Editores. (Primera edición publicada en 1971)

Gharibyan, H., & Gunsaulus, S. (2006). Gender gap in computer science does not exist in one former soviet republic: results of a study. In *Proceedings of the 11th annual SIGCSE conference on Innovation and technology in computer science education* (pp. 222–226).

Giannini, T., & Bowen, J. P. (2017). Life in code and digits: when Shannon met Turing. *Electronic Visualisation and the Arts (EVA 2017),* 51–58.

Gilens, M., & Page, B. I. (2014). Testing theories of american politics: Elites, interest groups, and average citizens. *Perspectives on Politics, 12*(3), 564–581.

Gill, P. (2016). *16 Famous People Whose Talents Were Only Recognized After Their Deaths.* `https://www.scoopwhoop.com/`

People-Talent-Only-Recognized-After-Death/. (Accessed: 8-11-2020)

Global Recruiters of Palm Beach. (2016). *Creating High Performance Committed Teams.* https://youtu.be/MhzePUOcD_s. (Interview with Mario Moussa. Accessed: 18-10-2020)

Goertzel, B. (2014). Artificial general intelligence: concept, state of the art, and future prospects. *Journal of Artificial General Intelligence, 5*(1), 1–48.

Goldstein, T. R. (2011). Correlations among social-cognitive skills in adolescents involved in acting or arts classes. *Mind, Brain, and Education, 5*(2), 97–103.

Gomez, E. (2019). *Women in Artificial Intelligence: mitigating the gender bias.* https://ec.europa.eu/jrc/communities/en/community/humaint/news/women-artificial-intelligence-mitigating-gender-bias. (Accessed: 18-11-2020)

Goodfellow, I., Bengio, Y., Courville, A., & Bengio, Y. (2016). *Deep learning* (Vol. 1). Cambridge: MIT Press.

Goodfellow, I. J., Bengio, Y., & Courville, A. (2016). *Deep Learning.* London, England: MIT Press.

Google Books Ngram Viewer. (n.d.). *Google Books Ngram Viewer.* https://books.google.com/ngrams. (Accessed: 12-11-2020)

Google Colab. (2019). *Google Colab.* https://colab.research.google.com/.

Gottfredson, L. (1997). Mainstream science on intelligence: An editorial with 52 signatories, history, and bibliography. *Intelligence, 24*, 13–23. Retrieved from https://doi.org/10.1016/S0160-2896(97)90011-8 doi: 10.1016/S0160-2896(97)90011-8

Gray, J. (2008). *Perros de paja: Reflexiones sobre humanos y otros animales.* Barcelona, Buenos Aires, México. Paidós: Paidós. (Traducción de Albino Santos Mosquera. Primera Edición de Bolsillo)

Greene, V. (2001). Personal hygiene and life expectancy improvements since 1850: Historic and epidemiologic associations. *American Journal of Infection Control*, *29*(4), 203–206. Retrieved from `http://www.sciencedirect.com/science/article/pii/S0196655301826730` doi: https://doi.org/10.1067/mic.2001.115686

Grimée, E. (2018). *Ensemble Moxos*.

Guillopé, C., & Roy, M.-F. (Eds.). (2020). *A Global Approach to the Gender Gap in Mathematical, Computing, and Natural Sciences: How to Measure It, How to Reduce It?* `https://zenodo.org/record/3882609`.

Gunamuktananda, D. (2014). *Consciousness – the final frontier.* `https://youtu.be/loOX2ZdElQ4`. TEDx Talks. (Accessed: 28-9-2020)

Haier, R. J. (2016). *The neuroscience of intelligence.* New York, NY: Cambridge University Press.

Harari, Y. N. (2016). *Homo deus: A brief history of tomorrow.* Random House.

Hawkins, A. (2019). *Toyota will offer robot taxi rides during the 2020 Summer Olympics in Tokyo.* `https://www.theverge.com/2019/10/24/20930227/toyota-self-driving-car-taxi-tokyo-test-2020-olympics`. (Accessed: 6-11-2020)

Hawksworth, J., Berriman, R., & Cameron, E. (2018). *Will robots really steal our jobs?* `https://www.pwc.com/hu/hu/kiadvanyok/assets/pdf/impact_of_automation_on_jobs.pdf`. PricewaterhouseCoopers. (Accessed: 17-6-2020)

Helliwell, J. F., Layard, R., & Sachs, J. D. (2019). *World Happiness Report.* `https://s3.amazonaws.com/happiness-report/2019/WHR19.pdf`. (Accessed: 26-11-2020)

Hennekam, S., & Bennett, D. (2017). Sexual harassment in the creative industries: Tolerance, culture and the need for change. *Gender, Work & Organization*, *24*(4), 417–434.

Hernández, H. (2020). *Modelo de pacto de socios.* `https://asesorias.com/empresas/modelos-plantillas/`

pacto-socios/. (Accessed: 26-10-2020)

Herrmann, E., Call, J., Hernández-Lloreda, M. V., Hare, B., & Tomasello, M. (2007). Humans have evolved specialized skills of social cognition: The cultural intelligence hypothesis. *science*, *317*(5843), 1360–1366.

Hidalgo, C. (2019). *Una idea osada para reemplazar a los políticos — César Hidalgo.* https://youtu.be/UUp39T3fPAo. TED. (Accessed: 8-11-2020)

High-Level Expert Group on Artificial Intelligence. (2019). *Ethics guidelines for trustworthy AI.* https://ec.europa.eu/digital-single-market/en/news/ethics-guidelines -trustworthy-ai. (Accessed: 3-11-2019)

Hu, X., Choi, K., Lee, J. H., Laplante, A., Hao, Y., Cunningham, S. J., & Downie, J. S. (2016). WiMIR: An informetric study on women authors in ISMIR. In *International Society for Music Information Retrieval (ISMIR) Conference, 2016.*

Hubel, D. H., & Wiesel, T. N. (1962). Receptive fields, binocular interaction and functional architecture in the cat's visual cortex. *The Journal of Physiology, 160*(1), 106.

Human Brain Project. (2013). *Human Brain Project.* https://www.humanbrainproject.eu/en/science/overview/. (Accessed: 26-9-2020)

Huyer, S. (2015). *Is the gender gap narrowing in science and engineering.* https://en.unesco.org/sites/default/files/usr15_is_the_gender_gap_narrowing_in_science _and_engineering.pdf. (Accessed: 5-9-2020)

Hyland, P. K., Lee, R. A., & Mills, M. J. (2015). Mindfulness at work: A new approach to improving individual and organizational performance. *Industrial and Organizational Psychology, 8*(4), 576–602.

Ingels, B. (2019). *Floating cities, the LEGO House and other architectural forms of the future.* https://youtu.be/ieSV8-isy3M ?t=570. TED. (Accessed: 6-6-2020)

Innovation at Work. (2018). *Who's responsible for an autonomous vehicle accident?* https://innovationatwork.ieee.org/whos-responsible-for-an-autonomous-vehicle-accident/. (Accessed: 5-10-2020)

Interesting Engineering. (2020). *The world's first autonomous bricklaying robot can build a three bed, two bath home in under three days.* https://twitter.com/IntEngineering/status/1246086322802044928. (Accessed: 6-6-2020)

International Federation of Robotics. (2017). *Robot density rises globally.* https://ifr.org/ifr-press-releases/news/robot-density-rises-globally. (Accessed: 14-10-2020)

ISMIR. (n.d.). *ISMIR Home.* http://ismir.net/. (Accessed: 1-12-2020)

Itatí, A. (2006). El acceso de las mujeres a la educación universitaria. *Revista argentina de sociología, 4*(7), 11–46.

Ito, E. (2020). *8-year-old violin prodigy pulling all the right strings.* http://www.asahi.com/ajw/articles/AJ202001220003.html. (Accessed: 5-11-2020)

ITU. (2019). *New ITU data reveal growing Internet uptake but a widening digital gender divide.* https://www.itu.int/en/mediacentre/Pages/2019-PR19.aspx. (Accessed: 8-11-2020)

Jao, N. (2018). *WeChat now has over 1 billion active monthly users worldwide.* https://technode.com/2018/03/05/wechat-1-billion-users/. (Accessed: 8-11-2020)

Jing, M., & Lee, A. (2017). *Where is China's Silicon Valley?* https://www.scmp.com/tech/start-ups/article/2106494/where-chinas-silicon-valley. (Accessed: 26-10-2020)

Johnson, H. M. (1911). Clever hans (the horse of mr. von Osten): A contribution to experimental, animal, and human psychology. *The Journal of Philosophy, Psychology and Scientific Methods, 8*(24), 663–666. (Pfungst, Oskar. With an introduction (and four supplements) by C. Stumpf. Translated by Carl L. Rahn. With a prefatory note by James R. Angell.)

Johnson, S. K., Hekman, D. R., & Chan, E. T. (2016). If there's only one woman in your candidate pool, there's statistically no chance she'll be hired. *Harvard Business Review, 26*(04), 1–7.

Kansaneläkelaitos–The Social Insurance Institution of Finland. (2020). *Results of Finland's basic income experiment: small employment effects, better perceived economic security and mental wellbeing.* https://www.kela.fi/web/en/news-archive/-/asset_publisher/1N08GY2nIrZo/content/results-of-the-basic-income-experiment-small-employment-effects-better-perceived-economic-security-and-mental-wellbeing. (Accessed: 6-11-2020)

Kapur, S. (2006, 02). From underdogs to tigers: The rise and growth of the software industry in Brazil, China, India, Ireland, and Israel. *The Economic Journal, 116*(509), F156-F157. Retrieved from https://doi.org/10.1111/j.1468-0297.2006.01069_1.x doi: 10.1111/j.1468-0297.2006.01069_1.x

Keightley, D. N. (1985). *Sources of Shang history: The oracle-bone inscriptions of bronze age China.* Berkeley MA: University of California Press.

Kelly, L. (2017). *Prophets, prophecy, and oracles in the roman empire: Jewish, christian, and greco-roman cultures.* Routledge.

Keras. (2020). *Using Pre-Trained Models.* https://keras.rstudio.com/articles/applications.html. (Accessed: 14-10-2020)

Kirschniak, C. (2018). *Impact of artificial intelligence in Germany.* https://www.pwc.de/de/business-analytics/sizing-the-price.pdf. (Accessed: 17-6-2019)

Klaric, J. (2017). *Un crimen llamado educación.* https://youtu.be/7fERXOOXAIY. Jürgen Klaric. (Accessed: 23-10-2020)

Klein, L., & Pradinaud, J. (2012). *Mark Zuckerberg: The real face behind Facebook.* STP Productions. (Motion Picture, Klein

(Director), Pradinaud (Producer))

Ko, C.-R., & An, J.-I. (2019). Success factors of student startups in korea: From employment measures to market success. *Asian Journal of Innovation & Policy, 8*(1).

Koch, C. (2019). *Consciousness and its place in nature.* `https://youtu.be/_rcPtU1n1WM`. Todd Boyle. (Access 26-9-2020)

Kuncheva, L. (2004). *Combining pattern classifiers: Methods and algorithms.* London: John Wiley & Sons.

Kuncheva, L. (2016). *Reviewing revolution...* `https://lucykuncheva.co.uk/review_revolution.html`. (Accessed: 13-11-2020)

LaCiudaddelasIdeas. (2017). *Debate de Inteligencia Artificial (IA) — CDI 2016.* `https://youtu.be/UNCUkklG4II`. (Accessed: 24-2-2020)

LaGamma, A. (2000). *Art and oracle: African art and rituals of divination.* New York: Metropolitan Museum of Art. (With an essay by John Pemberton III)

Lamech, R., & Saeed, K. (2003). *What international investors look for when investing in developing countries.* `http://documents1.worldbank.org/curated/es/272751468762026583/pdf/280910InvestorsPaper0EMS0No6.pdf`. The World Bank Group. (Paper No. 6. Accessed: 10-6-2020)

LandingAI. (2020). *AI transformation playbook: How to lead your company into the AI era.* `https://landing.ai/wp-content/uploads/2020/05/LandingAI_Transformation_Playbook_11-19.pdf`. (Accessed: 8-7-2020)

Lapuschkin, S., Wäldchen, S., Binder, A., Montavon, G., Samek, W., & Müller, K.-R. (2019). Unmasking clever Hans predictors and assessing what machines really learn. *Nature Communications, 10*(1), 1–8.

Lara, C. (2020). Cambio climático, movimientos sociales y políticas públicas: una vinculación necesaria. In J. C. Postigo (Ed.), *Consejo latinoamericano de ciencias sociales* (pp. 11–14).

http://bibliotecadigital.ciren.cl/bitstream/handle/
123456789/29265/363.7387POS2013.pdf?sequence=
1&isAllowed=y.

LeCun, Y., Boser, B. E., Denker, J. S., Henderson, D., Howard, R. E., Hubbard, W. E., & Jackel, L. D. (1990). Handwritten digit recognition with a back-propagation network. In *Advances in neural information processing systems* (pp. 396–404).

LeCun, Y., Cortes, C., & Burges, C. (1998). *The MNIST database*. http://yann.lecun.com/exdb/mnist/. (Accessed: 6-6-2020)

Lee, H. R., & Šabanović, S. (2014). Culturally variable preferences for robot design and use in South Korea, Turkey, and the United States. In *2014 9th ACM/IEEE International Conference on Human-Robot Interaction (HRI)* (pp. 17–24).

Lee, K.-F. (2018). *AI super-powers*. New York: Houghton Mifflin Harcourt.

Liang, F., Gotham, M., Johnson, M., & Shotton, J. (2017). Automatic stylistic composition of Bach chorales with deep LSTM. In *Proceedings of the 18th International Society for Music Information Retrieval Conference (ISMIR-17), Suzhou, China* (pp. 449–456). Suzhou, China: International Society for Music Information Retrieval.

Liu, P. J., Saleh, M., Pot, E., Goodrich, B., Sepassi, R., Kaiser, L., & Shazeer, N. (2018). Generating Wikipedia by summarizing long sequences. *arXiv preprint arXiv:1801.10198*.

Livni, E., & Koft, D. (2017). *The decline of the large US family, in charts*. https://qz.com/1099800/average-size-of-a-us-family-from-1850-to-the-present/. (Accessed: 9-11-2020)

Loscocco, K., & Bird, S. R. (2012). Gendered paths: Why women lag behind men in small business success. *Work and Occupations*, *39*(2), 183–219.

Lovelace, A. (1843). Notes on L. Menabrea's Sketch of the Analytical Engine Invented by Charles Babbage, Esq. *Taylor's Scientific Memoirs*, *3*, 666-731.

Luftkin, B. (2020). *What the world can learn from Japan's robots.* `https://www.bbc.com/worklife/article/20200205 -what-the-world-can-learn-from-japans-robots`. (Accessed: 6-6-2020)

Machine Learning News. (n.d.). *Machine learning news.* `https:// groups.google.com/g/ml-news`. (Accessed: 14-11-2020)

Makridakis, S. (1995). The forthcoming information revolution: Its impact on society and firms. *Futures*, *27*(8), 799–821.

Makridakis, S. (2017). The forthcoming Artificial Intelligence (AI) revolution: Its impact on society and firms. *Futures*, *90*, 46–60.

Maldonado Villafuerte, R., Puerta Gonzalez, J. A., & Araujo, I. (2014). *Ensamble Moxos.* `https://youtu.be/4UXaAqCIUng`. Aizpuru, Luis. (Accessed: 25-2-2020)

Mallett, J. D., & Schroeder, J. L. (2015). Academic achievement outcomes: A comparison of montessori and non-montessori public elementary school students. *Journal of Elementary Education*, *25*(1), 39–53.

Mańdziuk, J., & Żychowski, A. (2019). DeepIQ: A human-inspired AI system for solving IQ test problems. In *2019 International Joint Conference on Neural Networks (IJCNN)* (pp. 1–8). IEEE.

Mantha, Y., & Hudson, S. (2018). *Estimating the gender ratio of AI researchers around the world.* `https://medium.com/ element-ai-research-lab/estimating-the-gender-ratio -of-ai-researchers-around-the-world-81d2b8dbe9c3`. (Accessed 8-6-2020)

Marčelja, S. (1980). Mathematical description of the responses of simple cortical cells. *Journal of Neuophysiology*, *70*(11), 1297–1300.

Marks, D. F. (2010). IQ variations across time, race, and nationality: An artifact of differences in literacy skills. *Psychological Reports*, *106*(3), 643–664. Retrieved from `https://doi.org/10.2466/pr0.106.3.643-664` (PMID: 20712152) doi: 10.2466/pr0.106.3.643-664

Martin, G. (2011). *Gabriel garcía marquez. una vida*. Debate.

Martinho-Truswell, E., Miller, H., Asare, I.-N., Petheram, A., Stirling, R., Gómez-Mont, C., & Martínez, C. (2018). *Hacia una Estrategia de IA en México: Aprovechando la Revolución de la IA*. `https://www.gobiernohabil.com/2018/09/rumbo-una-estrategia-de-inteligencia.html`. (Accessed: 10-11-2019)

Massaro, D. W., & Anderson, N. H. (1971). Judgmental model of the ebbinghaus illusion. *Journal of experimental psychology*, *89*(1), 147.

McDermott, D. (1976). Artificial intelligence meets natural stupidity. *Acm Sigart Bulletin*(57), 4–9.

Melograni, P. (2007). *Wolfgang Amadeus Mozart: A biography*. Chicago & London: University of Chicago Press.

Miller, A. (2019). *Can machines be more creative than humans?* `https://www.theguardian.com/technology/2019/mar/04/can-machines-be-more-creative-than-humans`. (Accessed: 10-14-2020)

Montenegro Castedo, M., & Schulmeyer, M. (2018). Diferencia en la Formación y Desarrollo Laboral de Hombres y Mujeres en Áreas de Ingeniería en Bolivia. *Revista Aportes de la Comunicación y la Cultura*(24), 31–38.

Moravec, H. (1988). *Mind children: The future of robot and human intelligence*. Harvard University Press.

More, M. (2013). The philosophy of transhumanism. *The transhumanist reader*, *8*.

Moussa, M., Boyer, M., & Newberry, D. (2016). *Committed teams: Three steps to inspiring passion and performance*. John Wiley & Sons.

Müller, V. C., & Bostrom, N. (2016). Future progress in artificial intelligence: A survey of expert opinion. In *Fundamental issues of artificial intelligence* (pp. 555–572). Springer.

Muoio, D. (2016). *These 29 gorgeous images created by Google's AI raised almost $100,000 at auction.* `https://www.businessinsider.com/google-ai-images-raise-100000-at-auction-2016-2?r=US&IR=T&IR=T.` (Accessed: 12-2-2019.)

Nagpal, S., Singh, M., Singh, R., & Vatsa, M. (2019). Deep learning for face recognition: Pride or prejudiced? *arXiv preprint arXiv:1904.01219.* https://arxiv.org/pdf/1904.01219.pdf.

Navajo, J. (2013). *La importancia del Pacto de Socios en una startup.* `https://youtu.be/yYE0-ni1aJE.` Itnig. (Accessed: 26-10-2020)

Nechung–The State Oracle of Tibet. (1998). `https://web.archive.org/web/20061205234136/http://www.tibet.com/Buddhism/nechung_hh.html.` (Accessed: 8-11-2020)

NeurIPS. (2020). *Diversity & Inclusion.* `https://nips.cc/public/DiversityInclusion.` (Accessed: 14-12-2020)

Ng, A. (2017). *The state of artificial intelligence.* `https://youtu.be/NKpuX_yzdYs.` The Artificial Intelligence Channel. (Accessed: 12-2-2019)

Niebuhr, O., Tegtmeier, S., & Schweisfurth, T. (2019). Female speakers benefit more than male speakers from prosodic charisma training. A before-after analysis of 12-weeks and 4-h courses. *Frontiers in Communication, 4*, 12.

Nilsson, N. J. (2010). *The quest for artificial intelligence: A history of ideas and achievements.* New York, USA: Cambridge University Press.

Noor, N., Shapira, A., Edri, R., Gal, I., Wertheim, L., & Dvir, T. (2019). 3d printing of personalized thick and perfusable cardiac patches and hearts. *Advanced Science, 6*(11), 1900344. https://onlinelibrary.wiley.com/doi/pdf/10.1002/advs.201900344.

Nunez, M. (2018). *Artistas muy famosos reconocidos después de su muerte*. https://www.elsoldecuernavaca.com.mx/analisis/artistas-muy-famosos-reconocidos-despues-de-su-muerte-1372742.html. (Accessed: 8-11-2020)

OECD. (2020). *Unemployment rate (indicator)*. https://data.oecd.org/unemp/unemployment-rate.htm. (Accessed: 17-7-2020)

O'flaherty, W. D. (1986). *Dreams, illusion, and other realities*. Chicago: University of Chicago Press.

Olson, P. (2014). *Exclusive: The rags-to-riches tale of how Jan Koum built WhatsApp into Facebook's new $19 billion baby*. https://www.forbes.com/sites/parmyolson/2014/02/19/exclusive-inside-story-how-jan-koum-built-whatsapp-into-facebooks-new-19-billion-baby/#25c921552fa1. (Accessed: 26-10-2020)

O'Neill, J. (2001). Building better global economic brics. *Global Economics*.

Opinión. (2018). *Lavanderas Manos frías y corazón caliente*. https://www.opinion.com.bo/articulo/cochabamba/lavanderas-manos-fr-iacute-coraz-oacute-n-caliente/20180624235500617696.html. (Accessed: 23-10-2020)

Palacios, A. (2016). ¿qué es realmente el humanismo? *Revista Global, 65*. http://revista.global/que-es-realmente-el-humanismo/. (Accessed: 24-6-2021)

Pan, C. (2020). *China's Baidu finishes building 'world's largest' test ground for autonomous vehicle, smart driving systems*. https://www.scmp.com/tech/enterprises/article/3086353/chinas-baidu-finishes-building-worlds-largest-test-ground. (Accessed 20 September 2020)

Parra, A., Mendes, J. R., Valero, P., & Ubillús, M. V. (2016). Mathematics education in multilingual contexts for the indigenous population in Latin America. In *Mathematics education and language diversity* (pp. 67–84). Springer.

Pasquali, M. (2019). *Latin American startups—Statistics & Facts*. `https://www.statista.com/topics/4786/startups-in-latin-america/`. (Accessed: 19-10-2020)

Paul, E. S., & Kaufman, S. B. (2014). *The philosophy of creativity: New essays*. New York, USA: Oxford University Press.

Payne, A. (2020). *Germany is set to trial a Universal Basic Income scheme*. `https://www.weforum.org/agenda/2020/08/germany-universal-basic-income-trial-citizens/`. (Accessed: 5-10-2020)

PCMag Staff. (2009). *More Americans Go To Facebook Than MySpace*. `https://www.pcmag.com/archive/more-americans-go-to-facebook-than-myspace-241432`. (Accessed: 26-10-2020)

Perez, C. (2003). *Technological revolutions and financial capital: The dynamics of bubbles and golden ages*. Edward Elgar Publishing.

Perez, C. (2016). *Capitalism, technology and a green global golden age: The role of history in helping to shape the future*. United Kingdom. (Working paper)

Perez, C. (2018). *The fifth industrial revolution is happening— is it time to reshape our future?* `https://youtu.be/dhNd3tVR1hI`. Ellen MacArthur Foundation. (Accessed: 26-9-2020)

Perez, C. (2019). *Capitalism, technology and innovation*. `https://youtu.be/rE1oCzvo0E4`. UCL Institute for Innovation and Public Purpose. (Accessed: 6-6-2020)

Perna, R. (2017, 5). Intelligence: A measure of brainefficiency and the ability to integrate information. *PsycCRITIQUES*, *62*(19). (Article 4)

Petrie, A., & Petrie, J. (1986). *Mother Teresa*. Petrie Productions. (Motion Picture)

Piketty, T. (2014). *Capital in the twenty-first century* (A. Goldhammer, Trans.). Cambridge, Massachusetts and London: The

Belknap Press of Harvard University Press.

Pilon, M. (2015). *Monopoly's Inventor: The Progressive Who Didn't Pass Go?* https://www.nytimes.com/2015/02/15/business/behind-monopoly-an-inventor-who-didnt-pass-go.html. (Accessed: 6-11-2020)

Prego, C. (2020). *Científicas que conducían ambulancias en la guerra: Y otras mujeres en la ciencia.* Libros.com.

Press, G. (2018). *Why Facebook triumphed over all other social networks.* https://www.forbes.com/sites/gilpress/2018/04/08/why-facebook-triumphed-over-all-other-social-networks/#2861f0716e91. (Accessed: 22-10-2020)

PricewaterhouseCoopers. (2017). *Sizing the prize. What's the real value of AI for your business and how can you capitalise?* https://www.pwc.com/gx/en/issues/analytics/assets/pwc-ai-analysis-sizing-the-prize-report.pdf.

Prod'homme, J.-G. (1911). *Nicolo Paganini: A biography.* Fischer.

Prohorovs, A., Bistrova, J., & Ten, D. (2019). Startup success factors in the capital attraction stage: Founders' perspective. *Journal of East-West Business, 25*(1), 26–51.

Project Implicit. (2019). *Take a Test—Project Implicit - Harvard University.* https://implicit.harvard.edu/implicit/takeatest.html. (29-10-2020)

Proudfoot, K., Krieg, G., Rosen, J., Kohs, G., & Lee, C. (2017). *AlphaGo—The Movie — Full Documentary.* https://youtu.be/WXuK6gekU1Y. DeepMind. (Proudfoot, Krieg, y Rosen (Producción), Kohs (Dirección), Lee (Edición). Accessed: 6-11-2020)

Purdy, M., & Daugherty, P. (2016). *Why artificial intelligence is the future of growth.* https://www.accenture.com/t20170524t055435__w__/ca-en/_acnmedia/pdf-52/accenture-why-ai-is-the-future-of-growth.pdf. (Accesso: 12-10-2020)

Querejazú, V., Castellón, D. Z., & Córdova, J. M. (2015). *Global entrepreneurship monitor: Reporte nacional Bolivia 2014.* Escuela de la Producción y la Competitividad de la Universidad Católica Boliviana San Pablo.

Radio Télévision Suisse. (2011). *Il y a 40 ans, les femmes accèdent enfin au droit de vote.* https://web.archive.org/web/20110929153843/http://archives.tsr.ch/dossier-suffrage. (Accessed: 26-9-2020)

Raghuram, P., Herman, C., Ruiz-Ben, E., & Sondhi, G. (2017). *Women and IT scorecard-India. A survey of 55 firms spring 2017.* https://www.researchgate.net/profile/Gunjan_Sondhi/publication/314207187_Women_and_IT_Scorecard_-_India_A_survey_of_55_firms/links/58b9add645851591c5dbff5a/Women-and-IT-Scorecard-India-A-survey-of-55-firms.pdf. The Open University UK. (Accessed: 4-9-2020)

Raposo, V. L. (2019). The first Chinese edited babies: A leap of faith in science. *JBRA Assisted Reproduction, 23*(3), 197.

Redacción Economía y EFE. (2019). *Medellín inaugura primer centro para la Cuarta Revolución Industrial de América Latina.* https://www.elespectador.com/noticias/nacional/antioquia/medellin-inaugurara-primer-centro-para-la-cuarta-revolucion-industrial-articulo-852954. (Accessed: 26-2-2020)

Régner, I., Thinus-Blanc, C., Netter, A., Schmader, T., & Huguet, P. (2019). Committees with implicit biases promote fewer women when they do not believe gender bias exists. *Nature Human Behaviour, 3*(11), 1171–1179.

Reuters Staff. (2020). *George Floyd: America's racial inequality in numbers.* https://www.weforum.org/agenda/2020/06/george-floyd-america-racial-inequality/?utm_source=sfmc&utm_medium=email&utm_campaign=2721639_Agenda_weekly-5June2020&utm_term=&emailType=Newsletter. (Accessed: 6-6-2020)

Revista Capital. (2012). *Otra cosa es con Guital.* https://www.capital.cl/otra-cosa-es-con-guital/. (Accessed: 10-6-

2020)

Rio's carnival by the numbers. (2018). `https://www.france24`
`.com/en/20180209-rios-carnival-numbers`. (Accessed: 5-
11-2020)

Ritchie, G. (2007). Some empirical criteria for attributing creativity
to a computer program. *Minds and Machines*, *17*(1), 67–99.

Robert, L. P. (2019). Are automated vehicles safer than manually
driven cars? *AI & SOCIETY*, *34*(3), 687–688.

Rogers, M. (2016). *Cyber security & social media: How big is your
digital footprint and why should you care.* `https://docs.lib`
`.purdue.edu/dawnordoom/2016/presentations/3/`. Purdue
University. (DOI:10.5703/1288284316578. Accessed: 21-11-
2020)

Ronell, A. (2002). *Stupidity.* University of Illinois Press.

Ruiz, B. R., & Marín, R. R. (2012). *The struggle for female suffrage
in Europe: voting to become citizens.* The Netherlands, Leiden:
Brill.

Samuel, A. L. (1959). Some studies in machine learning using the
game of checkers. *IBM Journal of Research and Development*,
3(3), 210–229.

Sandberg, A., & Bostrom, N. (n.d.). *Whole brain emulation: A roadmap* (Tech. Rep. No. 2008?3). Oxford University: Future of Humanity Institute.
http://www.fhi.ox.ac.uk/Reports/2008-3.pdf.

Savery, J. R. (2015). Overview of problem-based learning: Definitions and distinctions. *Essential readings in problem-based
learning: Exploring and extending the legacy of Howard S. Bar-
rows*, *9*, 5–15.

Scharmen, F. (2019). *Jeff Bezos Dreams of a 1970s Future.*
`https://www.bloomberg.com/news/articles/2019-05-13/`
`why-jeff-bezos-s-space-habitats-already-feel-stale`.
(Accessed: 8-11-2020)

Schellenberg, E. G. (2004). Music lessons enhance IQ. *Psychological
Science*, *15*(8), 511–514.

Schmidhuber, J. (2015). Deep learning in neural networks: An overview. *Neural Networks, 61*, 85–117.

Schopenhauer, A. (2008). *El arte de tratar con las mujeres.* Alianza Editorial. (Collected and organized by Franco Volpi. Translation Fabio Morales García. Original work published 1944 and 1951.)

Schumpeter, J. A. (1934). *The theory of economic development (2017, with a new introduction by John e. Elliott).* New York, Roudledge.

Schwab, K. (2016). *The Fourth Industrial Revolution.* New York: Crown Business.

Scudder, J. (2015). *Astroquizzical: Is There Gold On Mars?* `https://www.forbes.com/sites/jillianscudder/` `2015/10/04/astroquizzical-is-there-gold-on-mars/` `#5e4508f029a5`. Forbes. (Accessed: 27-8-2020)

Shelley, M. (1818/1979). *Frankenstein or the Modern Prometheus.* New York: Bantam Pathfinder.

Silver, D., Huang, A., Maddison, C. J., Guez, A., Sifre, L., Van Den Driessche, G., ... Hassabis, D. (2016). Mastering the game of go with deep neural networks and tree search. *Nature, 529*(7587), 484–489.

Skolkovo Innovation Center. (2019). *Annual Report 2019.* `https://www.yumpu.com/en/document/read/62821242/` `sk-annual-report-2019-eng-web-2`. (Accessed: 20-10-2020)

Slim, C. (2016). *Sólo con empleo y educación se resuelve pobreza.* `https://youtu.be/twmcVTIyWPQ?t=2802`. carloslimvideoficial. (Accessed: 8-6-2020)

Smith, A. (1776/2010). *The wealth of nations: An inquiry into the nature and causes of the wealth of nations.* Harriman House Limited.

Solomon, M. (1990). *Beethoven essays.* Harvard University Press.

Solomon, M. (1995). *Mozart: A life* (1st ed.). New York: Harper-Collins.

SpaceX. (2020). *SpaceX.* `https://www.spacex.com/human
-spaceflight/mars/`. (Accessed: 6-6-2020)

Stacy, S. (2018). *Best Public Datasets for Machine
Learning and Data Science.* `https://medium.com/
towards-artificial-intelligence/the-50-best-public
-datasets-for-machine-learning-d80e9f030279`. (Accessed: 14-10-2020)

StartupChile. (2010). *Startup Chile.* `https://www.startupchile
.org`.

Statista. (2019). *Number of monthly active twitter users
worldwide from 1st quarter 2010 to 1st quarter 2019.*
`https://www.statista.com/statistics/282087/
number-of-monthly-active-twitter-users/`. (Accessed:
18-10-2020)

Statista. (2020a). *Number of annual active consumers across
Alibaba's online shopping properties from 2nd quarter
2015 to 2nd quarter 2020.* `https://www.statista.com/
statistics/226927/alibaba-cumulative-active-online
-buyers-taobao-tmall/`. (Accessed: 18-10-2020)

Statista. (2020b). *Number of monthly active facebook users worldwide as of 2nd quarter 2020.* `https://www.statista.com/
statistics/264810/number-of-monthly-active-facebook
-users-worldwide/`. (Accessed: 18-10-2020)

Statista. (2020c). *Number of monthly active WeChat
users from 2nd quarter 2011 to 1st quarter 2020.*
`https://www.statista.com/statistics/255778/
number-of-active-wechat-messenger-accounts/`. (Accessed: 18-10-2020)

Steele, K. M., Bass, K. E., & Crook, M. D. (1999). The mystery of
the Mozart effect: Failure to replicate. *Psychological Science,*
10(4), 366–369.

Sturm, B. L. (2014). A simple method to determine if a music
information retrieval system is a horse. *IEEE Transactions on
Multimedia, 16*(6), 1636–1644.

Suton, R. S., & Barto, A. G. (2018). *Reinforcement learning: An introduction* (2nd ed.). Cambridge MA, USA: MIT Press.

Sánchez, S. (2015). *Alucinantes nanorobots combatirán el cáncer navegando por nuestras venas.* `https://youtu.be/XLkfqiuOvqQ`. El Futuro Es Apasionante de Vodafone. (Accessed: 6-11-2020)

Tapia, A. (2019). *Talent framework, The Inclusive Leader, Optimizing diversity by leveraging the power of inclusion.* https://www.kornferry.com/content/dam/kornferry/docs/article-migration/Korn-Ferry-The-Inclusive-Leader_2019_06.pdf. Korn Ferry Institute. (Accessed: 2-2-2023)

The Economic Times. (2020). *Business News¿ Bill Gates.* `https://economictimes.indiatimes.com/news/bill-gates`. (Accessed: 8-11-2020)

The IEEE Global Initiative. (2019). *Ethically aligned design, first edition overview, a vision for prioritizing human well-being with autonomous and intelligent systems.* `https://ethicsinaction.ieee.org/`.

The Organisation for Economic Co-operation and Development. (2019). *OECD Principles on AI.* `https://www.oecd.org/going-digital/ai/principles/`.

Thomson, A., & Sharman, N. (2015). *Calculating Ada [Motion Picture].* `https://youtu.be/QgUVrzkQgds`. (Thomson (Producción) & Sharman (Dirección). Presentado por Hannah Fry. Accessed: 12-10-2020)

Tian, N., Kuimova, A., da Silva, D. L., Wezeman, P. D., & t. Wezeman, S. (2020). *Trends in world military expenditure, 2019.*

Toxopeus, J., & Sinclair, B. J. (2018). Mechanisms underlying insect freeze tolerance. *Biological Reviews*, *93*(4), 1891–1914.

Treffert, D. A. (2009). The savant syndrome: an extraordinary condition. a synopsis: past, present, future. *Philosophical Transactions of the Royal Society B: Biological Sciences*, *364*(1522), 1351–1357.

Turing, A. (1950). Computing machinery and intelligence. *Mind*, 433–460.

UCLA Mindful Awareness Research Center. (2018). *UCLA Mindful Awareness Research Center.* https://www.uclahealth.org/marc/mindful-meditations. (Accessed: 2-11-2020)

UNESCO Institute for Statistics. (n.d.). *UNESCO Institute for Statistics.* http://data.uis.unesco.org/Index.aspx?DataSetCode=EDULIT_DS#. (Accessed: 11-9-2020)

UNESCO Institute for Statistics. (2018). *Women in Science.* http://uis.unesco.org/sites/default/files/documents/fs51-women-in-science-2018-en.pdf. (Accessed: 5-9-2020)

UNESCO Institute for Statistics. (2019). *Women in Science.* http://uis.unesco.org/sites/default/files/documents/fs55-women-in-science-2019-en.pdf. (Accessed: 5-9-2020)

United Nations Educational, Scientific and Cultural Organization. (2006). *El carnaval de Oruro. Inscrito en 2008 (3.COM) en la Lista Representativa del Patrimonio Cultural Inmaterial de la Humanidad (originalmente proclamado en 2001).* https://ich.unesco.org/es/RL/el-carnaval-de-oruro-00003. (Accessed: 5-11-2020)

United States Census Bureau. (2019). *Number of People With Master's and Doctoral Degrees Doubles Since 2000.* https://www.census.gov/library/stories/2019/02/number-of-people-with-masters-and-phd-degrees-double-since-2000.html. (Accessed: 28-10-2020)

Valcárcel, A. (2007). Vindicación del humanismo (xv conferencias aranguren). *Isegoría*(36), 7–61.

Vance, S. M., Bird, E. C., & Tiffin, D. J. (2019). Autonomous airliners anytime soon? *International Journal of Aviation, Aeronautics, and Aerospace*, *6*(4), 12.

Varma, R. (2009). Why I chose computer science? Women in India. *AMCIS 2009 Proceedings*, 413.

Velarde, G. (2019). Artificial intelligence and its impact on the fourth industrial revolution. *International Journal of Artificial Intelligence & Applications (IJAIA), 10*(6), 41-48. Retrieved from `http://aircconline.com/abstract/ijaia/v10n6/10619ijaia04.html`

Velarde, G. (2020a). *A 4.0 artificial intelligence strategy in Bolivia.* Duesseldorf: PRICA.

Velarde, G. (2020b). *Curso virtual de Inteligencia Artificial.* `http://blog/curso_virtual_de_inteligencia_artificial.html`. (Accessed: 10-11-2020)

Velarde, G. (2020c). *Una estrategia 4.0 de inteligencia artificial en Bolivia.* Gissel Velarde.

Velarde, G. (2020d). *¿Por qué crees que pocas chicas estudian ingeniería o informática?* `https://www.facebook.com/photo/?fbid=169960681308840&set=g.344845209451512`. (Accessed: 5-9-2020)

Velarde, G., Chacón, C. C., Meredith, D., Weyde, T., & Grachten, M. (2018). Convolution-based classification of audio and symbolic representations of music. *Journal of New Music Research, 47*(3), 191-205. Retrieved from `https://doi.org/10.1080/09298215.2018.1458885` doi: 10.1080/09298215.2018.1458885

Vernon, B. (2016). *Qué escucha Michael Phelps en sus audífonos antes de competir?* `https://www.univision.com/musica/rio-2016/que-escucha-michael-phelps-en-sus-audifonos`. (Accessed: 5-11-2020)

VICE News. (2020). *India Is Becoming Its Own Silicon Valley.* `https://youtu.be/YHVNWtBuDVk`. (Corresponsal: Krishna Andavolu. Accessed: 19-10-2020)

Villacañas, B. (2001). De doctores y monstruos: la ciencia como transgresión en Dr. Faustus, Frankestein y Dr. Jekyll and Mr. Hyde. *Asclepio, 53*(1), 197–212.

Vincent, J. (2019). *Former Go champion beaten by DeepMind retires after declaring AI invincible.*

https://www.theverge.com/2019/11/27/20985260/
ai-go-alphago-lee-se-dol-retired-deepmind-defeat.
(Accessed: 6-11-2020)

Vise, D. (2017). *The Google story.* New York: Pan Books. (First published 2005)

Wachowski, L., Wachowski, L., & Silver, J. (1999). *Matrix.* Warner Bros. (Wachowski & Wachowski (Dirección), Silver (Producción))

Waldinger, R. (2016). *What makes a good life? Lessons from the longest study on happiness.* https://youtu.be/8KkKuTCFvzI. TED. (Accessed: 19-10-2020)

Walker, A. (2020). *Are self-driving cars safe for our cities?* https://www.curbed.com/2016/9/21/12991696/driverless-cars-safety-pros-cons. (Accessed: 6-11-2020)

Wang, H., Li, J., Li, W., Gao, C., & Wei, W. (2018). CRISPR twins: A condemnation from Chinese academic societies. *Nature, 564*(7736), 345–346.

Wikipedia. (2011). *Music Schools.* https://en.wikipedia.org/wiki/Music_school#cite_note-The_Great_Soviet_Encyclopedia_1979-9. (Accessed: 30-8-2020)

Wikipedia. (2013). *Problema del trigo y del tablero de ajedrez.* https://es.wikipedia.org/wiki/Problema_del_trigo_y_del_tablero_de_ajedrez#cite_note-4. (Accessed: 24-11-2020)

Wikipedia. (2019). *Manifestaciones y protestas en 2019.* https://es.wikipedia.org/wiki/Categor%C3%ADa:Manifestaciones_y_protestas_en_2019. (Accessed: 6-11-2020)

Wikipedia. (2020). *Lists of datasets for machine learning research.* https://en.wikipedia.org/wiki/List_of_datasets_for_machine-learning_research. (Accessed: 10-6-2020)

Wilde, G. (2010). Entre la duplicidad y el mestizaje: prácticas sonoras en las misiones jesuíticas de Sudamérica. *A tres bandas: mestizaje, sincretismo e hibridación en el espacio sonoro iberoamericano*, 103–112.

Winston, P. H. (1992). *Artificial intelligence*. United States of America: Addison-Weley.

Witten, I. H., Frank, E., & Hall, M. A. (2017). *Data mining: Practical machine learning tools and techniques* (4th ed.). Morgan Kaufmann.

Women in Machine Learning. (n.d.). *Women in Machine Learning*. https://groups.google.com/g/women-in-machine-learning. (Accessed: 14-11-2020)

Women in Music Information Retrieval. (2016). *Women in music information retrieval*. https://wimir.wordpress.com/mentoring-program/. (Accessed: 10-11-2020)

World Bank. (2010). *Reporte cambio climático*. http://documents1.worldbank.org/curated/es/985501468170350669/pdf/530770WDR00SPA00Box0361490B0PUBLIC0.pdf.

World Basic Income. (2020). *Basic income beyond borders: How a worldwide basic income could tackle global inequality and end extreme poverty.* https://www.ubi.org/_data/site/270/pg/28/WorldBasicIncome-Basicincomebeyondbordersv4.pdf. (Accessed: 6-11-2020)

World Economic Forum. (2018). *Asia works: The fourth industrial revolution.* https://youtu.be/-f_RwbfuwjU. (Accessed: 1-10-2020)

World Economic Forum. (2020a). *Diversity, Equity and Inclusion 4.0. A toolkit for leaders to accelerate social progress in the future of work.* https://www.weforum.org/reports/diversity-equity-and-inclusion-4-0-a-toolkit-for-leaders-to-accelerate-social-progress-in-the-future-of-work. (Accessed: 9-7-2020)

World Economic Forum. (2020b). *The global gender gap report 2020.* http://reports.weforum.org/global-gender

`-gap-report-2020/`. (Access 27-8-2020)

World Intellectual Property Organization. (2019). *WIPO Technology trends 2019. Artificial Intelligence.* `https://www.wipo.int/publications/en/details.jsp?id=4396`. Geneva: Author. (Accessed: 2-6-2020)

Worldometers. (2020). *Worldometers.* `https://www.worldometers.info`. (Accessed: 23-11-2020)

Wulf, A. (2019). *AI is incredibly smart, but it will never match human creativity.* `https://thenextweb.com/syndication/2019/01/02/ai-is-incredibly-smart-but-it-will-never-match-human-creativity/`. (Accessed: 14-10-2020)

Xomnia. (2014). *Impact with AI.* `https://www.xomnia.com`. (Accessed: 15-10-2020)

Yeager, D. S., Hanselman, P., Walton, G. M., Murray, J. S., Crosnoe, R., Muller, C., ... others (2019). A national experiment reveals where a growth mindset improves achievement. *Nature*, *573*(7774), 364–369.

Yoshimura. (2019). *Himari Yoshimura—7 yo Japan—1st Grand Prize—International Grumiaux Competition 2019—Paganini.* `https://youtu.be/YipD8Npugvg`. International Grumiaux Violin Competition. (Accessed: 5-11-2020)

Young, J. (2013). *Schopenhauer.* London and New York: Routledge.

Zanganehpour, S. (2018). *AI-supported global governance through bottom-up deliberation.* `https://globalchallenges.org/library-entries/ai-supported-global-governance-through-bottom-up-deliberation/`. (Accessed: 26-6-2020)

Zhou, S., Gordon, M., Krishna, R., Narcomey, A., Fei-Fei, L. F., & Bernstein, M. (2019). Hype: A benchmark for human eye perceptual evaluation of generative models. In *Advances in neural information processing systems* (pp. 3449–3461). (http://papers.nips.cc/paper/8605-hype-a-benchmark-for-human-eye-perceptual-evaluation-of-generative-models.pdf)

Zukin, S., & Papadantonakis, M. (2017). Hackathons as co-optation ritual: Socializing workers and institutionalizing innovation in the new economy. In *Precarious work* (pp. 157–181). Emerald Publishing Limited.

Appendix A

A.1 Trends in AI

As seen in Chapter 1, Table 1.1 shows that Deep Learning was the most used technique followed by multi-task learning and neural networks. Table 1.2 shows a summary of the most relevant functional applications. Computer vision filings reached just over 21 000, about four times the number of Natural Language Processing or Speech processing filings, and about ten times the number of robotics filings. Robotics had an average growth rate of 55 percent. Within computer vision, the preferred sub-categories were biometrics, with more than 6 000 filings, followed by character recognition, scene understanding, as well as image and video segmentation. Within Natural Language Processing (NLP), semantics had more than a thousand filings and grew by 33 percent annually. Sentiment analysis grew by 28 percent annually. Note this sub-category only represents one percent of NLP filings. As for speech processing, speech recognition and speaker recognition both grew by 12 percent annually, and speech-to-speech grew by 15 percent annually.

Transportation had the highest number of patent filings within application fields, with a total of 8 764 filings in 2016. The sub-category with the highest number of filings was autonomous vehicles, but aerospace/avionics had the highest average annual growth of 67 percent. Telecommunications also had a significant number of filings, with the sub-categories of radio and television broadcasting, and computer/internet networks standing out. Security also stood out for the number of filings in 2016, followed by medical and life sciences, as well as personal devices, computing, and human–computer interaction (HCI) with significant growth in affective computing, as shown in Table 1.3.

A.2 Survey: AI Trends

AI Trends

Considering Artificial Intelligence (AI) as the umbrella term for Machine Learning and Deep Learning, I would appreciate it if you take this 7-question survey (~5 minutes long). Your answers will be handled anonymously and will be used for research purposes. Thank you in advance! Kind regards, G. Velarde, Independent Researcher. `https://gvelarde.com/`

* Required

1. In your opinion, what are the Top 5 trends in AI research? Please, list them in order from (1) most relevant to (5) least relevant. *

Your answer

2. What should be the Top 5 priorities AI research? Please, list them in order from (1) most relevant to (5) least relevant. *

Your answer

3. In your opinion, what are the Top 3 most profitable applications of AI? Please, list them in order from (1) most profitable to (3) least relevant.

Your answer

4. Which option describes you best? *

○ OPTIMIST: AI will help us solve the most challenging world problems and will bring us closer to live in a world of unlimited wealth globally. In the future, we will enjoy the broad adoption of intelligent automation, and humans will work only on tasks of their preference.

○ PESSIMIST: AI could be our last invention. Artificial General Intelligence may occur in the future. Optimists underestimate the problems associated with superintelligence dominating humans. Since in the future, intelligent machines will take all important decisions for us (humans), we will be just a second class entity and many people will not be motivated to work.

○ PRAGMATIC: AI will potentiate economic growth where applied; it will increase labor productivity and will create new revenue streams on diverse areas. Some jobs will be lost, but more jobs will be created. Decision-making will increase its value and will remain as a particular task performed by people, not machines. The wealth gap may widen between those exploiting AI benefits and those who do not. Effective regulations will control AI and its dangers. Research on human intelligence augmentation will be fundamental.

○ DOUBTER: Artificial General Intelligence will never happen, such that AI will never outperform biological intelligence. Therefore, we should not consider AI as a threat to humans.

○ Other:

Where are you based? *
○ Africa
○ America
○ Asia
○ Australia/Oceania
○ Europe

---

6. What is your current role? *
○ Professor
○ Professional
○ Postdoc

○ PhD Student
○ Graduate student
○ Undergraduate student
○ Other:

7. What describes you best? *
○ Male
○ Female
○ Other:

Would you like to add a comment?
Your answer

Submit

Online AI trends surgey.
Data available at: http://gvelarde.com/a/data.html.

Example of an invitation to participate in the AI Trends survey, sent in 2020 to the channels of *Machine Learning News*, *Women in Machine Learning*, *ISMIR Community Announcements* y *Women in Music Information Retrieval*.

Subject: AI trends survey

Dear ML community,

please consider taking the following 7-question survey (∼5 minutes long) on AI trends. Your answers will be handled anonymously and will be used for research purposes.

The AI trends survey can be found here:
https://forms.gle/4GDJnXMNH7RKwonM9

Thank you in advance!

Kind regards,
Gissel Velarde
Researcher
www.gvelarde.com

A.3 Survey: AI's Impact

AI's impact

Please consider taking this short survey on the middle and long term impact of artificial intelligence (AI). Your answers will be handled anonymously and will be used for research purposes. Thank you in advance! Kind regards, G. Velarde, Independent Researcher. `https://gvelarde.com/`.

* Required

Which option describes you best? *

○ OPTIMIST: AI will help us solve the most challenging world problems and will bring us closer to live in a world of unlimited wealth globally. In the future, we will enjoy the broad adoption of intelligent automation, and humans will work only on tasks of their preference.

○ PESSIMIST: AI could be our last invention. Artificial General Intelligence may occur in the future. Optimists underestimate the problems associated with superintelligence dominating humans. Since in the future, intelligent machines will take all important decisions for us (humans), we will be just a second class entity and many people will not be motivated to work.

○ PRAGMATIC: AI will potentiate economic growth where applied; it will increase labor productivity and will create new revenue streams on diverse areas. Some jobs will be lost, but more jobs will be created. Decision-making will increase its value and will remain as a particular task performed by people, not machines. The wealth gap may widen between those exploiting AI benefits and those who do not. Effective regulations will control AI and its dangers. Research on human intelligence augmentation will be fundamental.

○ DOUBTER: Artificial General Intelligence will never happen, such that AI will never outperform biological intelligence. Therefore, we should not consider AI as a threat to humans.

○ Other:

Your country of residence: *
Your answer

Your nationality: *
Your answer

What is your current role? *
○ Professor
○ Professional
○ Postdoc
○ PhD Student
○ Graduate student
○ Undergraduate student
○ Other:

What is your highest academic degree completed? *
○ Ph.D.
○ Master
○ Bachelor
○ High school
○ Other:

How many years have you spent doing research in AI? *
○ 0
○ 1
○ 2
○ 3

○ 4
○ 5
○ between 6 and 9
○ between 10 and 19
○ ¿20

What is your age range? *
○ Under 21
○ Between 21-30
○ Between 31-40
○ Between 41-50
○ Between 51-60
○ Between 61-70
○ Above 71

What describes you best? *

○ Male
○ Female
○ Other:

Would you like to add a comment?
Your answer

Your answers will be handled anonymously and will be used for research purposes. The collected responses may be published. *
○ Agree

Submit

Online AI trends survey.
Data available at: http://gvelarde.com/a/data.html.

Example of an invitation to participate in AI's Impact survey, sent in 2020 to the channels of *Machine Learning News, Women in Machine Learning, ISMIR Community Announcements* y *Women in Music Information Retrieval.*

Subject: Share your opinion on the impact of IA, last call

Dear all,

I am writing a book contemplating the impact of IA. Those who have not yet given their opinion, are kindly invited to consider taking this short survey accessing the following Google form:

https://forms.gle/TMRRY6xfmeMSUsAa9

Thank you in advance!

Kind regards,

G. Velarde, PhD
Independent Researcher.
https://gvelarde.com/

A.4 Responses to the "Other" alternative to the question: *"Which option describes you best? *"*

Other: "AI (especially robotics), when well applied, has the potential to automate dangerous or tedious work, and to significantly increase the standard of living. AI *is*, however, just a technology and its positive and negative effects depend strongly on how it is used and regulated. Current trends point towards AI-related techniques (machine learning, pattern recognition, robotics etc.) being used to immensely enrich a small elite, exploit workers more efficiently (e.g. via the so-called gig economy), and support oppressive regimes all over the world. So I am rather pessimist about AI but not because of the technology itself but because of the economic and political system in which it will be deployed", a male professor.

Other: "IF we can solve the PESSIMIST problems then I am an OPTIMIST about the positive impact", a male postdoc.

Other: "AI will amplify sociological trends whatever those are", a female professional.

Other: "AI can potentiate economic growth where applied, but the majority of benefits will go to very few people. It will not increase labor productivity but will create new revenue streams while killing others. Many jobs will be lost, but some jobs will be created. The wealth gap will certainly widen between those corporations exploiting AI benefits and those that do not. Regulations might offer some control of the use of AI in certain spheres, but how effective those can be remains to be seen", a male professor.

Other: "Both PRAGMATIC and DOUBTER", a female PhD student.

Other: "I would describe myself as PRAGMATIC, with the difference that I do not believe that 'effective regulations will control AI and its dangers' ", a female postdoc.

Other: "I wanted to write 'Pragmatic' above with this comment: 1) Much depends on the nature and approach of people who use this. It can be mis-used. A parallel, I keep saying in these type of discussions is the use of atomic/nuclear power. It was mis-used during WWII. 2) I am not very sure about decisions being handled by humans; Reinforcement Learning (used by autonomous vehicles like drones, etc) is all about taking decisions", a male professor.

Other: "CRITIC. I identify most with the position of the Pragmatic, but I have a background in Neuroscience, and a nuanced understanding of how machine and human intelligence differ. My opinions on AI are closely interlinked with my skepticism/criticism of monopoly capitalism enjoyed by a handful of big tech companies", a female graduate student.

Other: "Close to PRAGMATIC, plus the fear that it becomes weapon (disinformation maker)", a female graduate student.

Other: "I think AI is already having positive and negative effects in different areas. I think this won't change in the future, there will new regulations but in other aspects there will be new harms. I also think that, with time, more people is conscience about the impact that AI can have", a male PhD student.

Other: "If we can solve challenges discussed above under 'PESSIMIST', then I think the future will look like that described above in 'OPTIMIST'! ", a male PhD student.

Other: "like OPTIMIST: AI future will growth together with business target!", a male professional.

Other: "Pragmatic, but I also realize the dark side of AI, especially when big tech companies are making money selling user's private information and using ads to influence their user's thoughts", a female PhD student.

Other: "Between optimist and pragmatic :)", a male postdoc.

Other: "somewhere between pessimist and pragmatic. I am often cynical in discussions around this, but I like to think there are plenty of pessimists out there who will help the AI future look like that of the pragmatic's", a female professional.

Other: "Pragmatic-Pessimist. It does not matter if AGI happens or not because we will empower machines to our own detriment anyway. Just as profit continues to override the environment and our humanity to each other, AI will be used in the service of the powerful", a male postdoc.

Other: " 'doubter' option does not work", a non-binary postdoc.

A.5 Isaac Asimov's Laws of Robotics

Isaac Asimov's Laws of Robotics (Nilsson, 2010, pp. 25):

- "Zeroth Law: A robot may not injure humanity, or, through inaction, allow humanity to come to harm".

- "First Law: A robot may not injure a human being, or, through inaction, allow a human being to come to harm".

- "Second Law: A robot must obey the orders given it by human beings except where such orders would conflict with the First Law".

- "Third Law: A robot must protect its own existence as long as such protection does not conflict with the First or Second Law".

Alphabetical Index

C

D

E

F

Facebook, 58, 59, 62, 73, 87

FaceMash, 59, 71

Fanfou, 60

features, 25

fixed mindset, *see also* growth mindset

Flores, Eugenio, 160

Flynn effect, 44

Flynn, James, 44

Fujifilm, 61

G

g-factor, 41–43

García Marques, Gabriel, 184

Gates, Bill, 59, 199

general intelligence, *see* intelligence

general intelligence factor, *see* g-factor

generative models, 49

genius, 41, 46, 47, 150

Go, 2, 15, 40

Goertzel, Ben, 40

gold, 57, 58

    new, 1, 56

Golgi, Camillo, 18

Google, 34, 58–62, 73

Graphic Processing Units, 19

Great Reset, 197

ground truth, 25

growth mindset, 148, 149

Guital, Marcelo, 90

H

Haier, Richard, 20, 32, 41

happiness, 168, 169

Harari, Yuval Noah, 139, 174

Hawking, Stephen, 174
Haydn, Joseph, 39
Hinton, Geoffrey, 19
Hubel, David, 18
Humanism, 164, 165
Humanist, 142

I
IBM, 58
Indonesia Investment Coordinating Board, 71
Industrial Revolution, *see* Revolution
Information Technology, 100
innovation, 55
Instagram, 62
intelligence, 15, 32, 46, 148, 181
    artificial, *see* Artificial Intelligence
    collective, 174
    definitions, 20
    general, 41–43, 45
    quotient, 41, 153
    tests, 41
intelligent
    automation, 2, 173, 174
    entities, 46
    machines, 25, 46, 181
IQ, *see* intelligence

J
Jacquard looms, 50
Jeopardy, 15
Jobs, Steve, 90
Joseph, Martin, 163
justice, 169

K
Kabat-Zinn, Jon, 162

Kaku, Michio, 175
Kant, Immanuel, 46
Klaric, Jürgen, 140
know-how, 70
Kodak, 61
Koller, Daphne, 75
Krizhevsky, Alex, 19
Kuncheva, Ludmila, 39
Kurzweil, Ray, 40, 174

L
label, 25
labor productivity, 173
Landing AI, 75
LeCun, Yann, 18, 73
Lee, Kai-Fu, 73
Lembong, Thomas, 71
literacy, 44
Lovelace, Ada, 17, 47, 50, 51, 183
Luddites, 51

M
Ma, Jack, 60
machine learning, 24
magical realism, 184
Makridakis, Spyros, 8
Maldonado, Raquel, 161
manipulation, 185, 200
Marić, Mileva, 109
Mars, 13
mathematical model, 24
Matrix, 174, 182
McDermott, Drew, 33
meditation, 182
Merkel, Angela, 13

Miebach, 97
MILA, 71
mindfulness meditation, 162, *see also* meditation
Minimun hardware, 81
monopoly, 195
Montessori, Maria, 141
Moravec, Hans, 18
Mother Teresa of Calcutta, 199
Mozart, Wolfgang Amadeus, 39, 150, 154
multi-planetary society, 13
Musk, Elon, 199
Myspace, 61

N
nanorobots, *see also* robot
natural disaster, 203
Naver, 73
neuromorphic computing, 183
neurons, 18, 183
new gold, 58, 185
New World Order, 56, 197
Ng, Andrew, 12, 75
Nilsson, Nils, 17, 21, 33, 184
noise, 39
novelty, 49
nutrition, 113

O
Offline-Pedia, 157
Open AI, 71, 175
open-source, 81

P
Paganini, Niccolò, 150
Page, Larry, 59, 61

virtue, 140, 168, 169
virtuosity, 150

W

Wachowski, Lana, 174
Wachowski, Lilly, 174
WeChat, 58, 73
WhatsApp, 62, 87
Wiesel, Torsten, 18
Wikipedia, 49
Winston, Patrick, 21
Wojcicki, Susan, 60
World Basic Income, *see also* Universal Basic Income

X

Xiaomi, 85
Xiaonei, 60, 73
Xomnia, 75

Y

Yahoo, 87
Yandex, 73
YouTube, 60

Z

Zhongguancun, 60
Zuckerberg, Mark, 59, 62, 199

About the Author

Dr. Gissel Velarde obtained her Ph.D. in Computer Science and Engineering from the prestigious Aalborg University, Denmark. She also holds a Master of Science in Electronic Systems and Engineering Management from the University of Applied Sciences Südwestfalen, Soest, Germany, and a Licenciatura in Systems Engineering from Universidad Católica Boliviana, La Paz. Velarde has over 20 years of experience in the field of Engineering and Computer Science, both in Academia and Industry. In 2020, she was named "Notable Woman" by the diversity promotion committee of The International Society for Music Information Retrieval. Among several honors, awards, and prizes, she received a teaching award in 2021 for her lecture and her students' learning outcomes on Selected Topics in Artificial Intelligence at Universidad Privada Boliviana.

Learn more about her at: http://gvelarde.com.